AF838116

HOBSON'S CHOICE

ENGLISH BRACKET CLOCK REPEATING WORK

THE WORKING SKETCHES MADE OVER MORE THAN SEVENTY YEARS
by

CHARLES HOBSON

Edited with additional Illustrations and Text
by

LAURENCE HARVEY and CHARLES ALLIX

**"HOBSON'S CHOICE"
AN APOLOGY**

The sketches shown on pages 7-11 of "Hobson's Choice" are the work and copyright of Daniel W Parkes. They will continue to appear by his generous permission. The five Parkes sketches have long been filed in the Archives of Charles Hobson together with his own. They were wrongly credited to the latter in good faith. Sincere apologies and thanks to Daniel Parkes.

It would (I think) be very frivolous, to speak of the various Contrivances, and Methods of Repeating-Work, and the Inventors of them; and therefore I shall say nothing of them.

(Derham, *The Artificial Clock-Maker*, 1696)

Charles H Hobson

Two thousand copies of this work have been printed.
This copy is No. 29

MALCOLM GARDNER SEVENOAKS

© Copyright 1982 Sketches Charles Hobson

**© Copyright 1982 Other Illustrations and Text
Laurence Harvey and Charles Allix**

World Copyright Reserved

ISBN 0 9508258 0 8

All rights reserved. No part of this publication may be reproduced, stored in a retrieval system, or transmitted in any form or by any means electronic, mechanical, photocopying, recording or otherwise without the prior permission in writing of the copyright owners.

Published by
Charles Allix trading as Malcolm Gardner,
Horological Bookseller & Publisher,
Bradbourne Farmhouse,
Sevenoaks, Kent.

Telephone: (0732) 451311

Printed in England by Lakeside Printing Ltd., Sevenoaks.

CONTENTS

INTRODUCTION

The publication of any worthwhile information regarding the design and operation of English silent pull and pull quarter repeating work has been negligible since Thiout. This is hardly surprising as until the early part of the present century the most varied and interesting feature of clockwork was notable mainly for its removal. With the formation of many important horological collections at the turn of the century there arose a new interest in the restoration of these much abused timekeepers.

It was at about this time that a son, Charles, was born to George and Louisa Hobson of Goswell Terrace, Clerkenwell. Although there were no clock or watchmakers in the immediate family, it has recently come to light that there is a direct male line back to John Hobson, Freeman of the Worshipful Company of Blacksmiths, who was one of the thirteen petitioners to Charles I in 1630 for the formation of a separate Company for clockmakers.

In 1907, thirteen-year old Charles Hobson went to Messrs. J.J. Elliott of Rosoman Street, Clerkenwell, where he was apprenticed to become a clockmaker (a much abused description now, but then a literal one) later going to work in the evenings with Daniel Parkes of Spencer Street. Mr. Parkes, the grandfather of the present Daniel, was one of the first restorers of old clocks as we understand the term today, maintaining many important collections. He naturally fired the young apprentice with enthusiasm for this field of horology, teaching him the fundamentals of early English clockwork.

Soon after the completion of Charles Hobson's apprenticeship, the first World War broke out. He served first as an E.R.A. in the Royal Navy, later volunteering for the newly formed Royal Naval Air Service in which he later became a Chief Mechanic. He was seriously injured in a flying accident. Happily, after many months of hospital treatment he was sufficiently recovered to take charge of a travelling workshop in the Middle East until the Armistice.

Upon his return to this country, Charles Hobson set up in a business at his home in Wynyatt Street, Clerkenwell, with his elder brother, a brilliant watchmaker, trading as Hobson Bros. This name he retained until his retirement even though his brother met an untimely death in 1927. Very soon after starting the business he moved to larger premises along the road, sharing them with the late Tom Miles and taking on two workmen. By 1928 Charles Hobson had moved to Walton Street in the West End, from where on occasion he worked on Royal clocks besides maintaining those in many large houses, institutions and collections. Seven years later he joined with Philip Clowes, Captain Jauncy, R.N. and Mr. James in a business at Beauchamp Place which flourished until 1939 when, for the second time, war intervened.

It was during this last period that he managed to acquire many historic tools from the estate of the Jump family. The best of these were a Sam. Nelson wheel engine, a large throw and a centring tool belonging originally to Benjamin Vulliamy and a large beam compass and spring winder of George Graham's. Philip Clowes purchased a large mahogany bench with Vulliamy's name engraved on a brass plate set into it. Sadly, the whereabouts of this is now unknown.

The Second World War saw Charles Hobson setting up and running a large workshop engaged in toolmaking for the Air Ministry and producing some highly important and secret work. Shortly before the end of hostilities he decided to move down to the South Coast and so purchased large premises in Hove from which he continued his business much as before although spending much of his time in London.

In 1963 at the age of 69, he decided to retire and moved to the nearby village of Henfield intending to work on his own for the first time and having a house and workshop built for the purpose. Naturally, he found himself more in demand than ever, as the swiftly rising value of early English clocks made expensive restorations economically viable. He had to refuse more work than he could undertake. In 1967 he was accorded the rare honour of being elected an Honorary Freeman of The Worshipful Company of Clockmakers in recognition of his contribution to horology over fifty years.

Laurence Harvey.
April 1982

ACKNOWLEDGEMENTS

Illustrations, other than Mr. Hobson's sketches, are by courtesy of:-

Guy Burnand Esq. 96 left, 101.
Clockmakers Company. 73 left and right, 83, 99.
British Museum. 28, 29, 35 right, 36, 37, 64 left and right. 75, 89.
Laurence Harvey, Esq. 78, 79.
Horological Workshops, Guildford. 69, 96 right, 100.
Beresford Hutchinson, Esq. 60.
Derek Roberts, Esq. 68 left and right.
T.E. Roxburgh, Esq. 47, 48, 49.
Sotheby, Parke Bernet & Co. 35 left.
Victoria & Albert Museum 93.
Anthony Woodburn, Esq. 34, left and right.

Michael Hurst, Esq. most kindly read the Preface.

PREFACE

ENGLISH SILENT PULL AND PULL QUARTER REPEATING WORK

English spring-clocks, as made from the beginning of the last quarter of the 17th century until late in the 18th, were very often furnished with repeating. This most ingenious and useful adjunct was provided mainly for use after dark and especially when the last light had been extinguished. By simply pulling a cord, the time shown by a clock could be sounded on bells to the nearest quarter of an hour. In an age of primitive lighting (not to mention the inordinate difficulty of igniting in the dark a sulphur "match" from a tinder-box in order to light a candle) the invention must have seemed to many who did not sleep well like the answer to a prayer. The only previous timetellers designed for the hours of darkness had been night clocks. Their rarity today suggests that few were made which did not end in flames.

Most English spring-clocks were designed to be portable. They were highly prized and were carried from room to room very much more than is now generally realised. Their tolerant verge escapements, light pendulum bobs and sensible top carrying handles well fitted them for duty by day and night. They could be put down on any level-ish surface and there would continue to go reasonably well even when much out of beat. Carried with care they would not stop while being moved. The invention of repeating naturally revolutionised the usefulness of such clocks. Even greater importance came to be attached to them. When later, presumably in the 19th century, the same clocks had their repeating torn out and discarded, it was a symptom of ignorance and incompetence as well as a sign of the times.

Many 17th century spring-clocks were timepieces. That is to say they had only a watch-part (going train) and did not strike. Others had both a watch-part and a clock-part (hour striking train and appurtenances) and struck the hours in passing. Yet others again struck in passing the hour at the hour and the quarter at the quarter. These were known as quarter clocks or three part clocks. At first, and before the invention of repeating, all English striking trains were controlled by count-wheels (locking plates). These advanced inexorably and therefore seemed to offer no possibility for repetition. The picture only changed in the reign of Charles II when in about 1676 Barlow and Quare produced clock repeating controlled by snails. From then onwards many spring-driven timepieces, hour striking clocks and quarter clocks, using the new invention, could at will be made to repeat hours and quarters. In timepieces the feature came to be known as silent pull and in hour striking clocks pull quarter. The required results were achieved in a number of different ways. Meanwhile, before the turn of the century, racks and snails superceded locking plates as controllers of striking work in most English clocks, whether spring or weight driven, bracket or longcase. Occasionally a transitional bracket clock will be found having hour striking on a locking plate; yet it will have a completely independent silent pull (see page 84). Silent pull and pull quarter clocks were soon followed by quarter and half-quarter repeating watches. These gradually superseded the bracket clock in the bedroom.

Mr. Hobson's sketches show the workings of various forms of silent pull and pull quarter. These are the most varied, interesting and least understood aspect of English clock-work.

The following are the main forms of repeating work as found in English spring-clocks:-

(1) **Timepieces with Silent Pull**

Here, when the cord is fully pulled and then released the last hour is first struck on the hour bell. It is followed by the last quarter sounded on a separate quarter bell or bells. The term "pull repeating" often used in connection with silent pull is something of a misnomer since the clock does not strike of its own accord. The silent pull is very ingenious. A subsidiary spring, cocked or wound by the pull only when required, drives a train of wheels known as the runners thereby sounding to the nearest quarter of an hour the time shown by the hands. The various ways in which this is achieved are explained alongside the sketches in relation to the actual examples shown. The earliest clocks, except for those made by Tompion, have no "all-or-nothing" piece. As a result a "short"pull, depending upon its length, will cause incorrect hours to be struck, or even quarters with no hours. The spring which drives a repeating train is sometimes a straight blade spring, cocked by the pull and then held by clickwork, and sometimes a miniature mainspring housed in a fixed-barrel and wound by pulling off a line wrapped round a deep pulley squared on to the barrel arbor. In general, blade springs are found mostly in striking clocks and coiled springs in timepieces. The reason is that silent pulls have more work to do than pull quarters. After about 1710, blade springs were used less and less as more quarter bells became fashionable. Tompion is only once known to have used a coiled spring in his repeating work (Symonds *Tompion*, fig. 205). In his "Mostyn" clock going for a year Tompion employed two "V"-shaped springs to drive his pull quarter. This form had long been used by locksmiths and gunmakers.

Many of the first table or bracket clocks with silent pull were beautifully small and light. This, coupled with the fact that they did not strike in passing, made them ideal for use upstairs. For this reason many people must have preferred them to hour striking clocks with pull quarter which were not only far heavier but also required setting to "not strike" at night if they were to hold their peace.

(2) **Hour Striking Clocks with Pull Quarter**

Here, when the cord is fully pulled and then released, the last quarter is sounded first on one or more quarter bells. Next the last hour is struck on the hour bell. In these clocks the rack striking work for the hours is already present. It requires only initially to be "warned" and subsequently released after the pulled-quarters, provided by a separate spring and wound only for the occasion, have been sounded. The striking train is designed not only to sound the hours in passing for a full eight days but also to allow for frequent repetition of the hours. There is then no danger of the striking part running down before the going part. In the 'Mostyn" clock not only will the going train run for 384 days but the striking train has the capacity to cover 420 days provided that the repeating work is not used. This allows 20.7 extra blows per day over a 365 day year. The clock was ordered by William III and is undoubtedly Tompion's masterpiece. It is mechanically-speaking the finest clock extant. Its making required incalculably greater precision of design and execution than any year-going weight driven clock. It certainly would not be conceived and made today.

(3) **Quarter Striking Clocks with Pull Repeating**

Here, when the cord is fully pulled and then released, the last quarter is sounded followed by the last hour. In this sort of clock both hour and quarter trains are already present (sometimes one train sounds both hours and quarters, then it is termed three parts in two). All that is necessary in an ordinary quarter clock in order to provide repeating is a means of sequential "warning" and release together with adequate resources to sustain the extra blows. The release in this case requires only a very short "pull" as no winding-up is involved.

Additional Notes

Since many people use the term "chime" when they mean "strike", a few remarks on terminology are desirable. A clock strikes the hours. It does not chime them. A quarter-clock is said to strike, sound or chime the quarters, the last term becoming more appropriate the more bells there are. Silent pull and quarter clocks should never be said to chime even though there may be many quarter bells.

The 18th century saw a gradual decline in popularity of clocks having silent pull and pull quarter. They were scarcely, if ever, made after about 1800. It is probable that the increasing availability of repeating watches, which were far more convenient for telling the time in the dark, was the main cause for the change in fashion. As the 19th century found its feet so lighting improved enormously. Both lamps and candles became very much brighter. Gas lighting began to appear in cities and towns after 1813, while from about 1831 matches became available which could be struck on the sides of their boxes. Clocks of all kinds, both English and imported, became ever more numerous with every house of any pretension having at least one in almost every room. Both French and English carriage clocks appeared, while Swiss repeating watches joined the English ones in the shops. In consequence bracket clocks were scarcely expected to be portable and were seldom if ever required to repeat. English clockmakers, however, still kept up a kind of pretence by providing bracket clocks (which by then were large and heavy, besides having anchor escapements with weighty pendulums) not only with carrying handles but also with pulls for repeating hours only. In these clocks the striking trains were debased (cheapened) so that their ability to provide many more blows than the 1248 required for eight days striking in passing was severely curtailed. If the pulls were used much the striking trains would run down long before the end of the week. This type of clock is said to have knock-up repeating as the hour rack hook is simply "knocked-up".

No aspect of English clockwork has ever suffered more at the hands of the repairers than the silent pull and pull quarter. Enough clocks survive in original unspoiled condition to establish that most of them were made in London. It is reasonable to assume that in their heyday there were enough workmen there who understood them. But what happened in the end when such a clock was sold to a home deep in the country? Even as late as 1820 the journey by coach from Newcastle-on-Tyne to London, for instance, occupied two whole days and the night between. Surtees wrote "I travelled by the old Highflyer coach, to catch which at eight o'clock in the morning I had to leave Hamsterley between five and six. Then, by a steady perservering grind continued all that day, all that night, and all the following day, we reached the dismal White Horse in Fetter Lane at eight that night. The fare was £6 inside, and it was considered very fine travelling". Thus, many a London clock must have been given out to a village clockmaker for cleaning and repair. Silent pulls and pull quarters were stoutly made; so why was the repeating so often stripped out and in what period? Was it done in the 19th century and if so what was the reason? Nobody of course knows. It may not be fanciful to suppose that the country man, on occasion, having dismantled such clocks could not then be bothered to overhaul and clean their many extra parts and instead discarded the repeating work. The times would have been in his favour. No doubt he would have said that this part of the clock was worn out. By then the need for repeating would anyhow have passed and perhaps owners no longer cared.

Towards the end of the 19th century the collecting of clocks suddenly became something of a vogue. It was then and since that repeating began to be reinstated. Few were able to do it, but conversely not much of this kind of work was required. "Grandfather Parkes" was one of those in London able to reproduce the missing parts as they should have been. He used un-restored clocks as his guide. There may have been one or two other workmen equally good. Others again were less scrupulous. They put into old clocks repeating bearing no resemblance to the original in either system or appearance. They used scarcely an old pair of holes. The customers, if they knew or minded, were in no position to object. This kind of work has continued up to the present day. On the other hand, in the past decade a new and altogether more serious type of person, often very competent and well trained in other branches of clockmaking, is striving to put back silent pull and pull quarter. It is for him that Mr. Hobson has generously allowed the publication of these records.

C.R.P. Allix

Timepiece with Silent Pull sounding the Quarters on a Single Bell

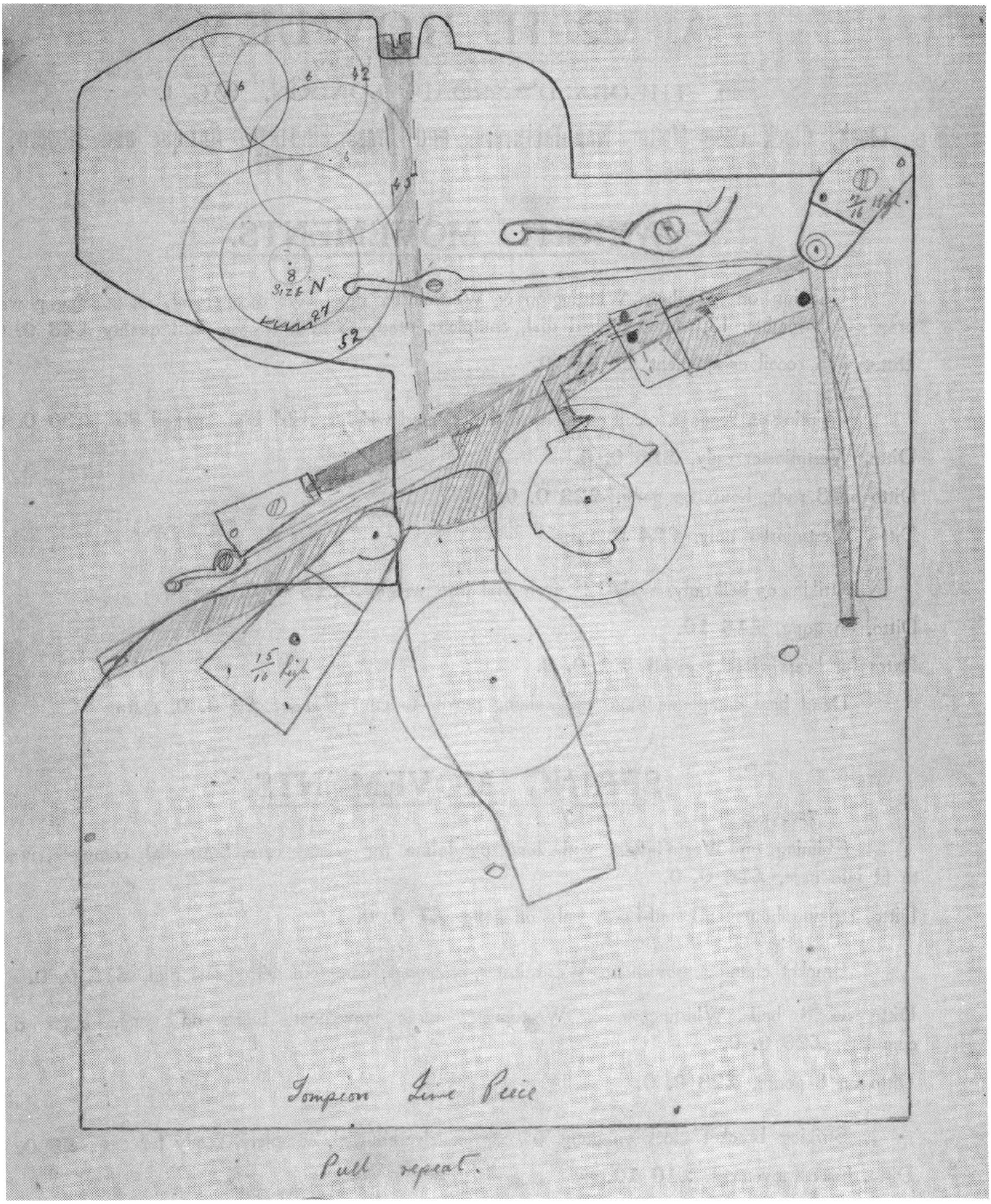

Tompion. The general arrangement of early silent pull. Tompion's system, as will be seen, is immensely complicated, employing some 50 individual parts where other makers managed with perhaps as few as 20. Nevertheless, Tompion's design is by far the best. It should work and feel like silk. Pages 8-11 show detail sketches of the same clock. Symonds *Tompion*, fig 114, shows a clock of this type complete in its case.

Timepiece with Silent Pull sounding the Quarters on a Single Bell

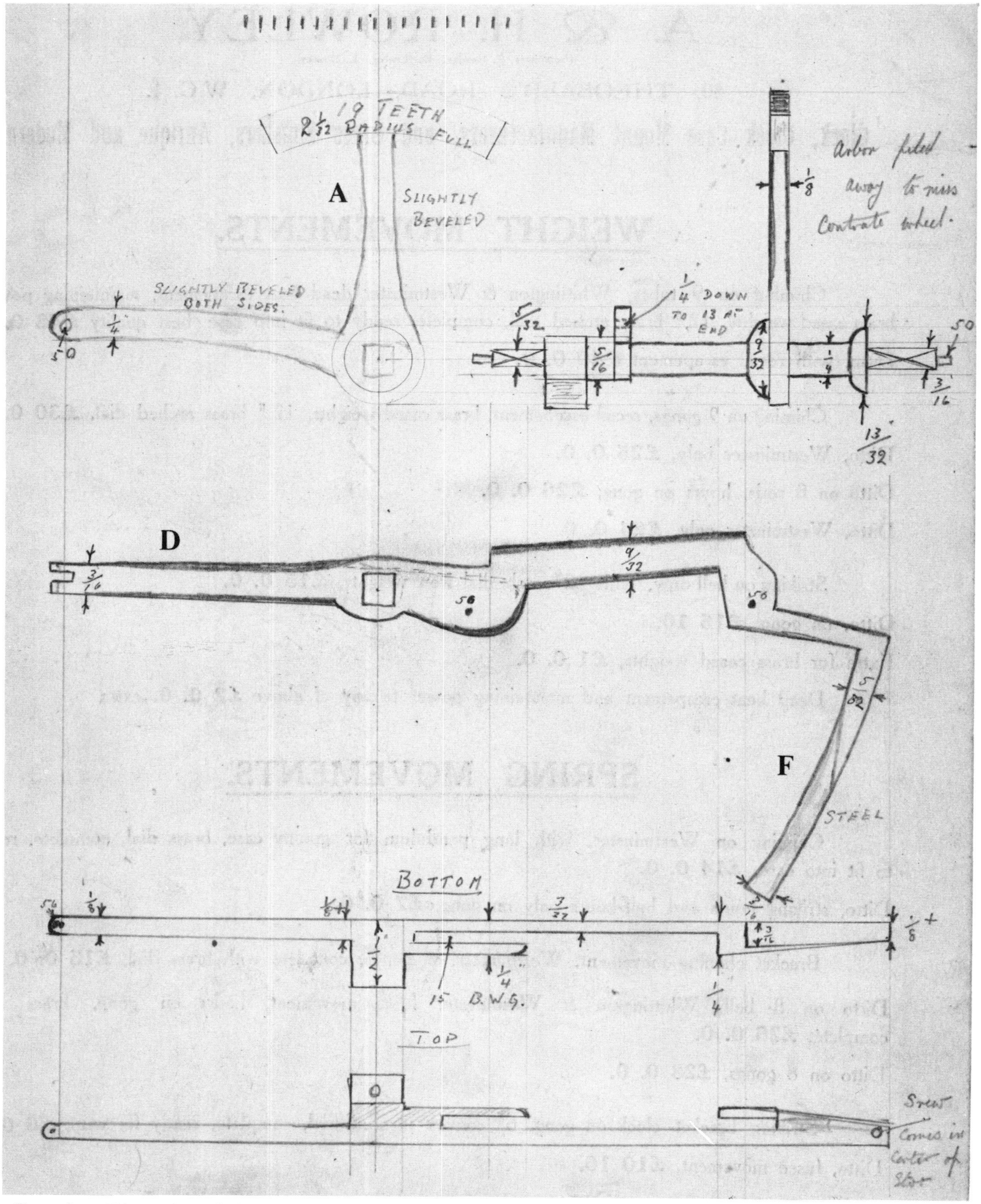

Tompion. Dimensioned sketches giving details of some of the parts shown in the layout of page 7. See also page 12.

Timepiece with Silent Pull sounding the Quarters on a Single Bell

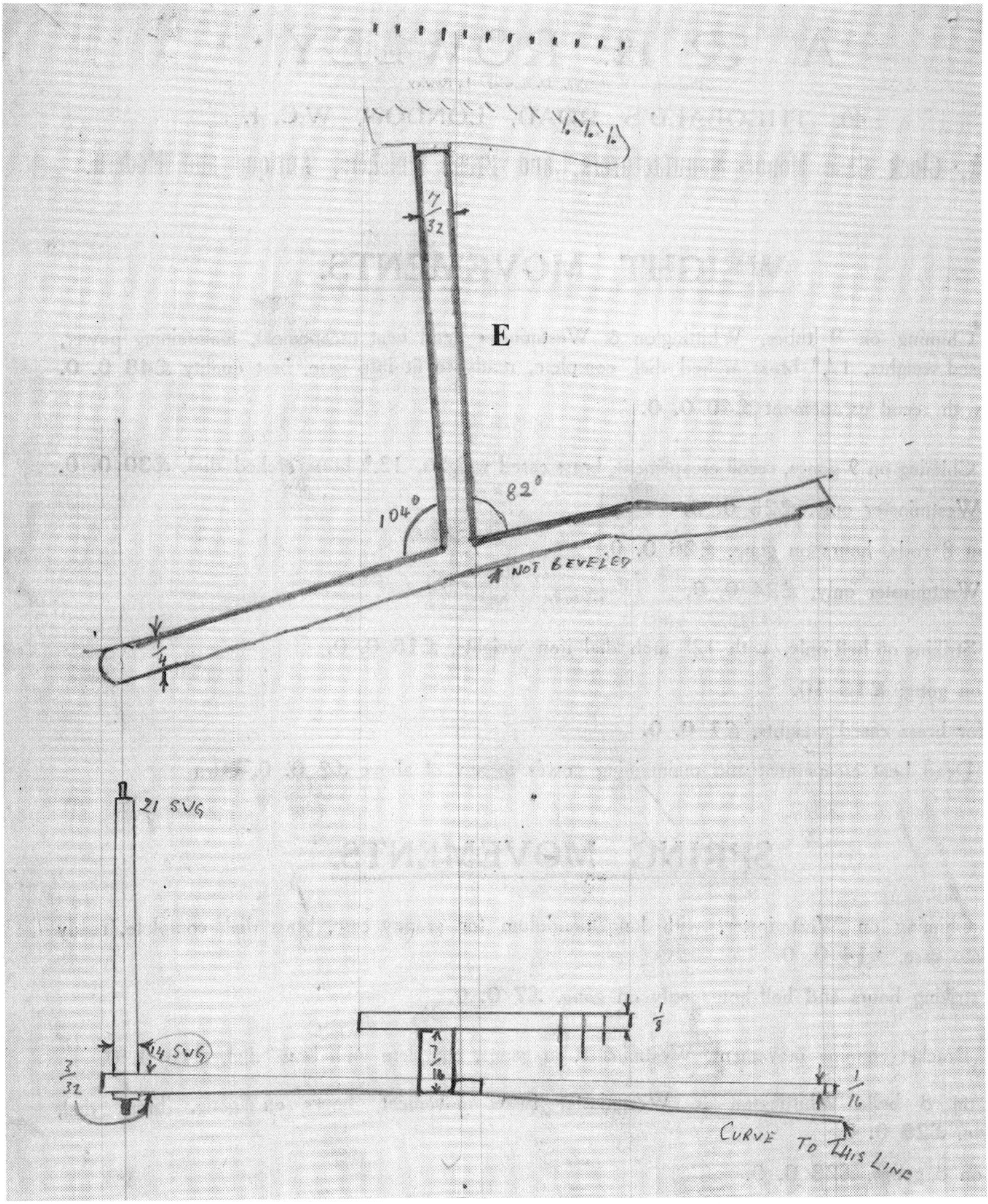

Tompion. Further sketches for use with pages 7 to 13.

Timepiece with Silent Pull sounding the Quarters on a Single Bell

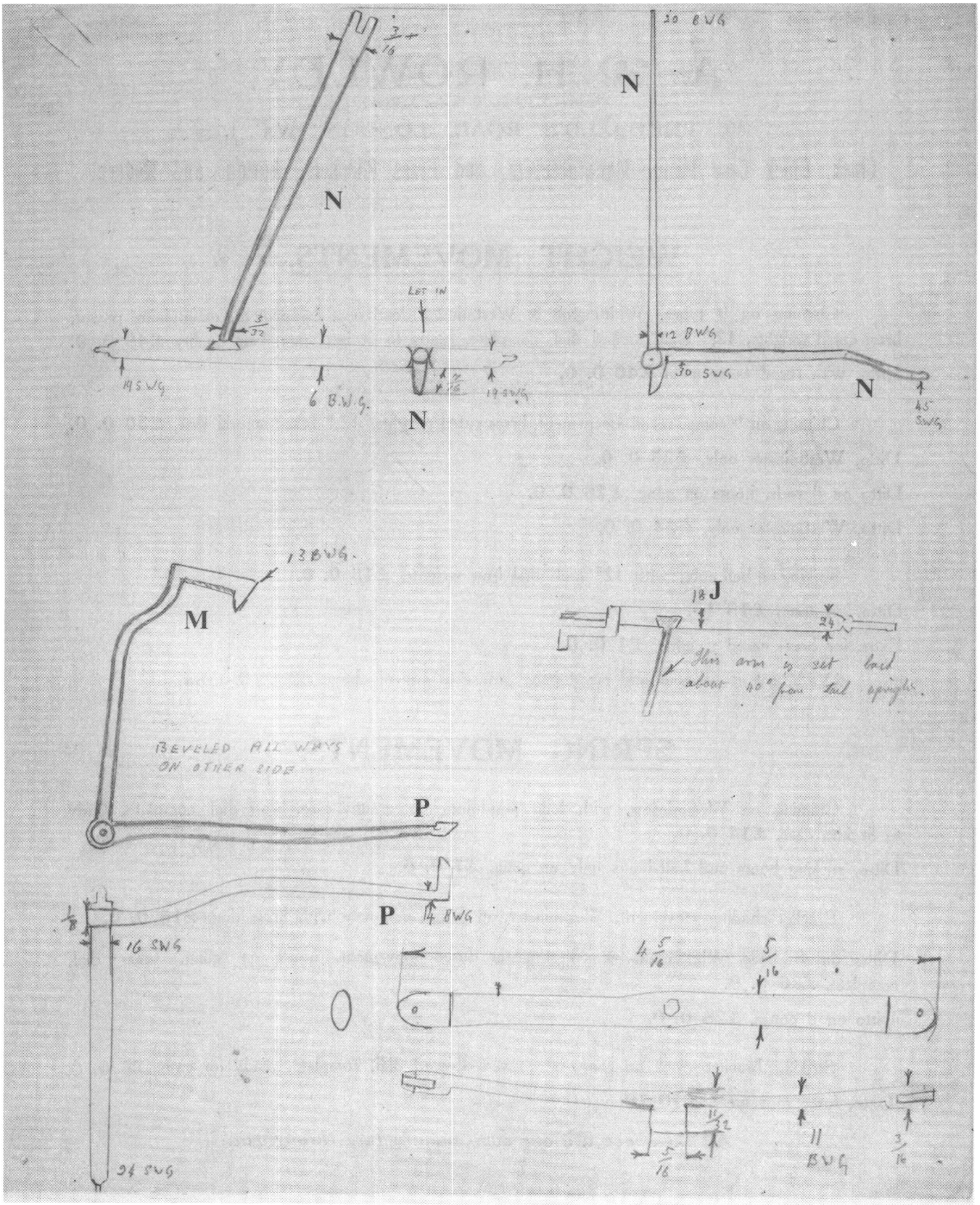

Tompion. Further sketches for use with pages 7 to 13.

Timepiece with Silent Pull sounding the Quarters on a Single Bell

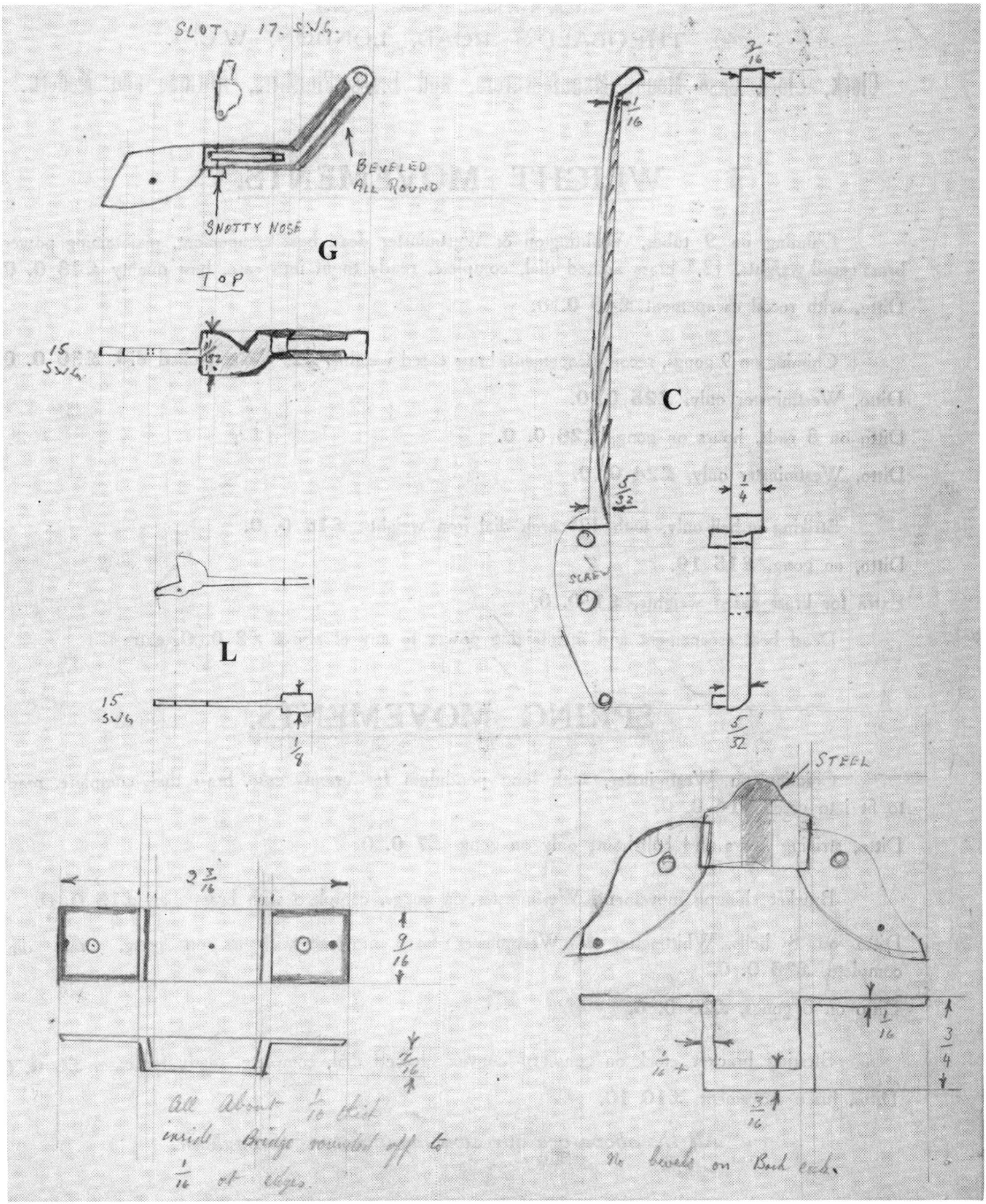

Tompion. Further sketches for use with pages 7 to 13.

Timepiece with Silent Pull sounding the Quarters on a Single Bell

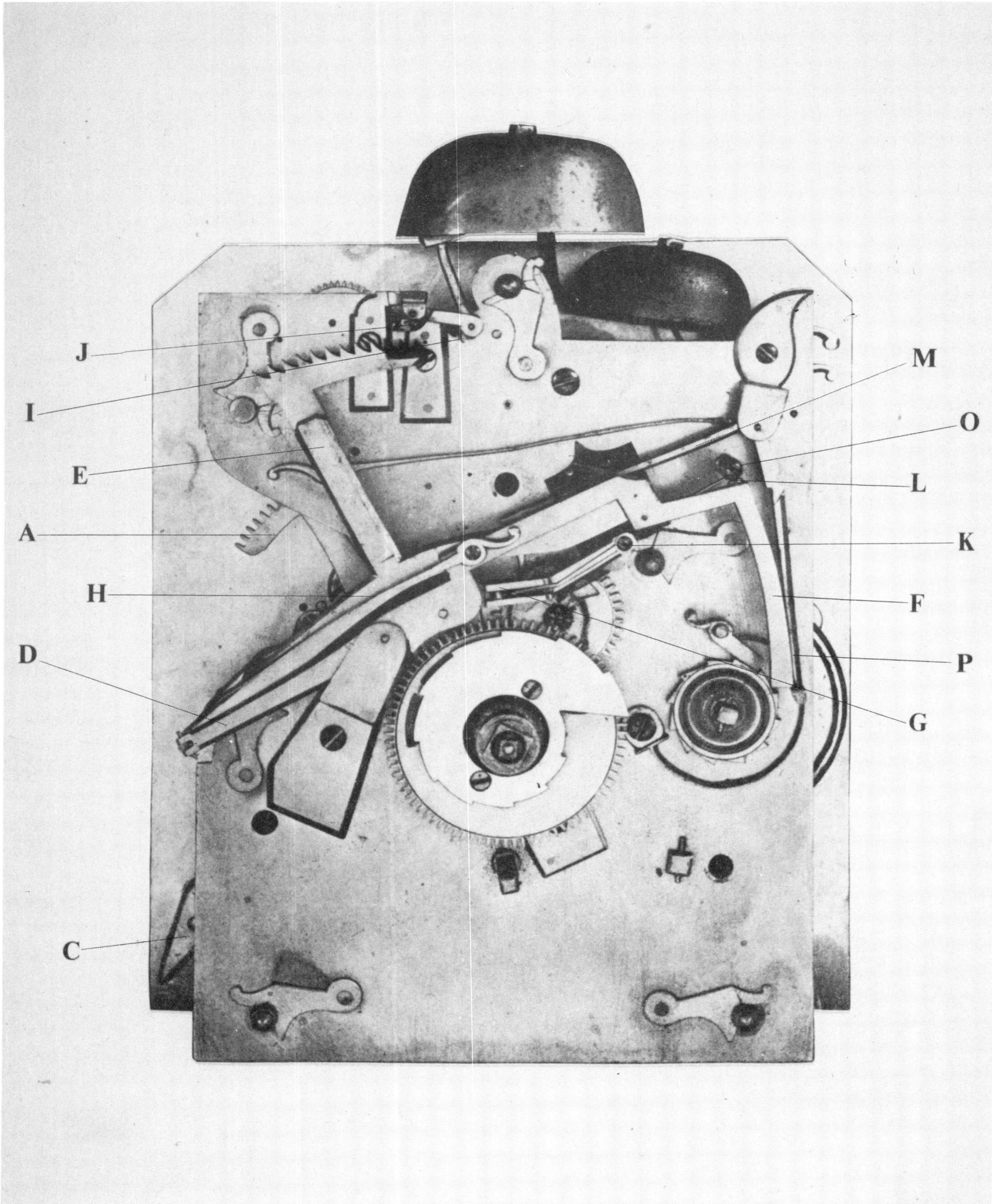

Tompion. Typical silent pull. See pages 7 — 13.

Timepiece Silent Pull — Description of the Action

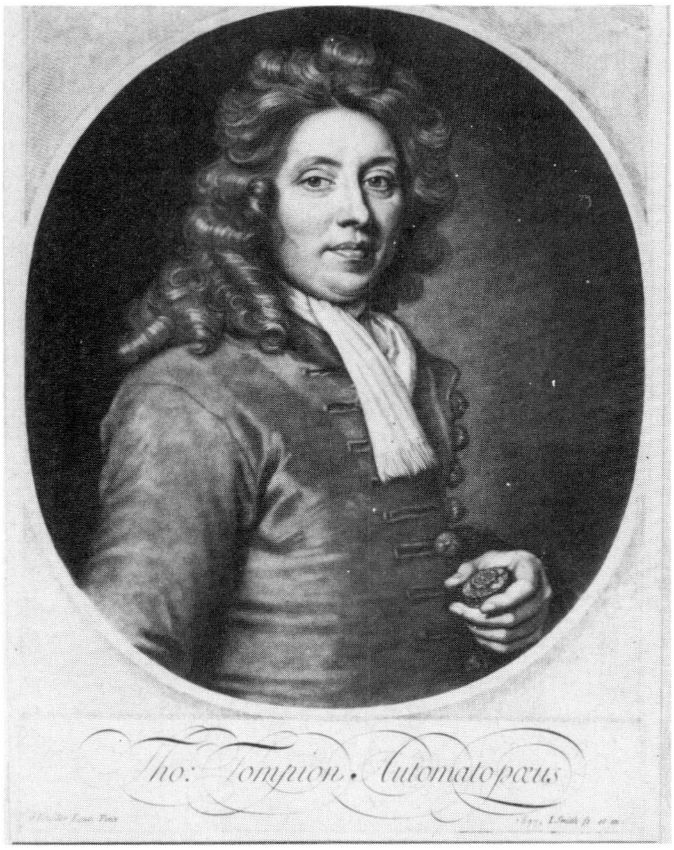

Tompion's silent pull, in common with all of his clock-work, is in a different "league" to that of any of his contemporaries. Despite the modern fashion for attempting to denigrate Tompion, educated examination of any of his work shows it to be of an altogether superior quality and refinement. It has about it a kind of "Rolls-Royce" subtlety in conception and execution difficult to express in words and yet readily discernable to anyone who wishes to see it.

Silent pull, as made by Tompion, may or may not have been the first type. As noted elsewhere, it is certainly by far the most complicated. It may be helpful to the reader to study the Knibb system (pages 36 and 37) before trying to understand Tompion's as shown pages 7-22.

In Tompion's silent pull (as indeed in all others) a completely independent train (the runners) is provided to strike the last hour followed by the last quarter and also to control the speed at which the blows are struck. An all-or-nothing piece K ensures that a too short pull, when released, will simply cause the runners to turn without any blows being struck. When the double-armed lever is pulled (see page 35 left) the internal motion rack A rotates clockwise turning pinion B (see page 17) thereby cocking the heavy blade spring C. The clickwork associated with the main wheel prevents the runners from being turned backwards during this operation. The large hour and quarter rack assembly DEF is carried on the arbor of the internal motion rack and thus also rotates clockwise on pulling off. The movement of both is checked when the rack tail G reaches a step on the hour snail. The rack tail incorporates for safety a sprung end. Part E of the rack assembly is pivoted at the lower end of D (see pages 8 to 11 for detail sketches) and is held down at this stage by means of a blade spring H so the teeth clear both the hour hammer tail I and the quarter hammer tail J during the pull off. As the rack tail (which is pivoted) reaches a step on the hour snail, the nose K rises, thereby raising the rack E which until now has been kept down by the spring H. The rack is maintained in its temporary raised position by the pivoted detent L which now tilts under the influence of a small spring and props up the squared end of the rack. The rising of the rack brings its teeth into the ambit of both hammer tails. When the pull is released and as the train runs, the hour teeth lift and release the hour hammer tail sounding the number of blows dictated by the hour snail. The quarters are achieved as follows. During the whole time that the clock is going the quarter snail carried on the minute wheel is followed by the tail of the lever M, which by means of a simple two-plane lever system NN (details page 10 top) continuously pumps the quarter hammer arbor (which has very long pivots and also a long end shake) backwards and forwards between the plates. A weak spring presses on the pump lever thus keeping the whole system in compression. After the hours have been struck, the quarter hammer tail, depending upon its lateral position, is raised and released by none, one, two or three of the three pins (or teeth) of graduated length on the side of the rack. On completion of the run, the raised detent L meets a fixed stop O thereby releasing rack E and allowing the spring H to restore it to its original lower position.

Further evidence of the subtlety of Tompion is to be found in his provision of a locking arm P on the quarter snail follower M. A high step on the quarter snail after the third quarter step moves the arm P into the path of rack assembly tail F thereby preventing pulling off during the hour changeover period of approximately four minutes.

Timepiece with Silent Pull sounding the Quarters on a Single Bell

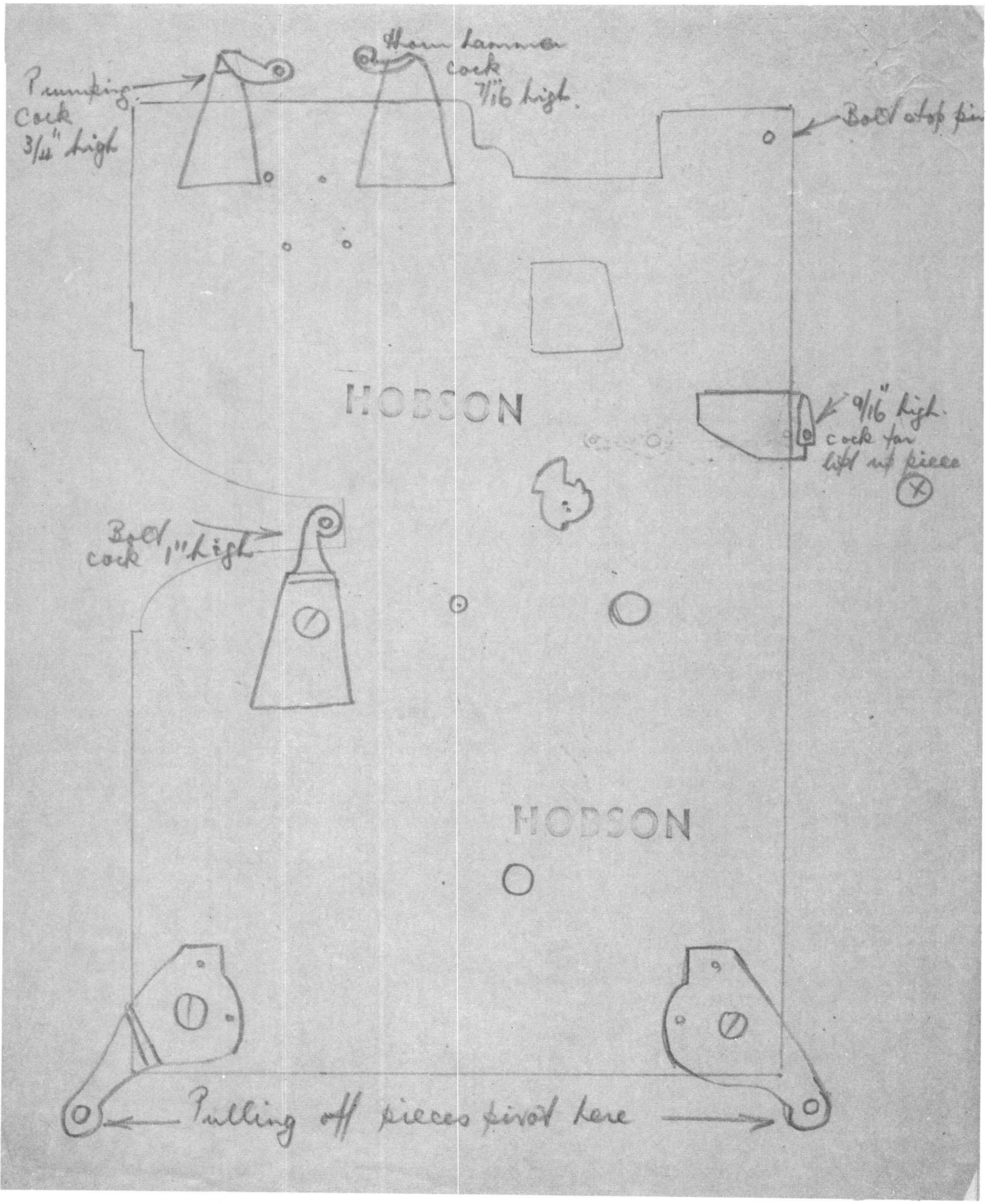

Tompion. Details of front plate of an earlier but rarer silent pull rather similar to that shown on pages 7 - 13. See also Symonds *Tompion*, fig. 174, for illustration of back plate of similar clock. This shows the pull eyes at rest in the slotted back plate. Compare with pulls shown on page 35 of Hobson.

Timepiece with Silent Pull sounding the Quarters on a Single Bell

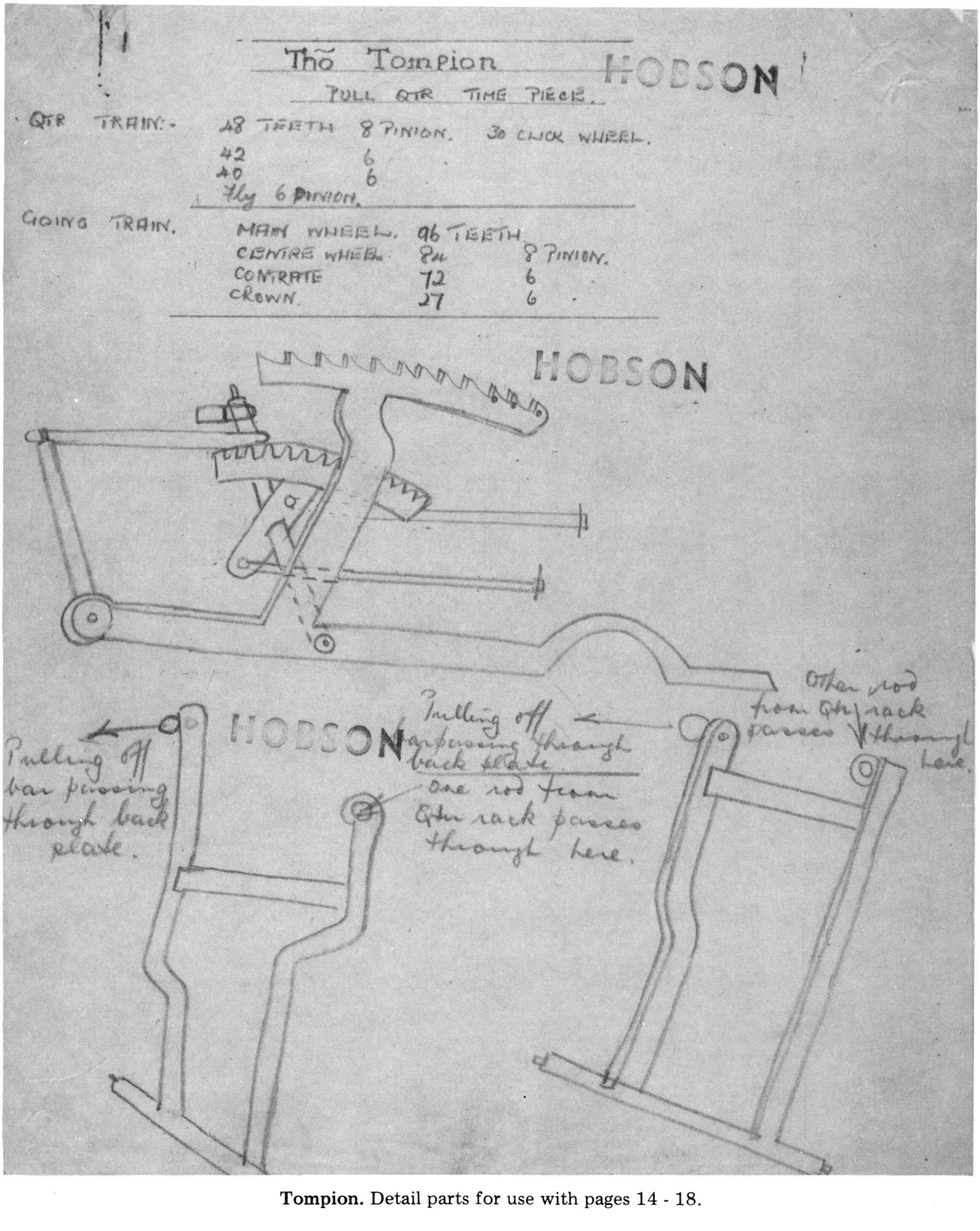

Tompion. Detail parts for use with pages 14 - 18.

Timepiece with Silent Pull sounding the Quarters on a Single Bell

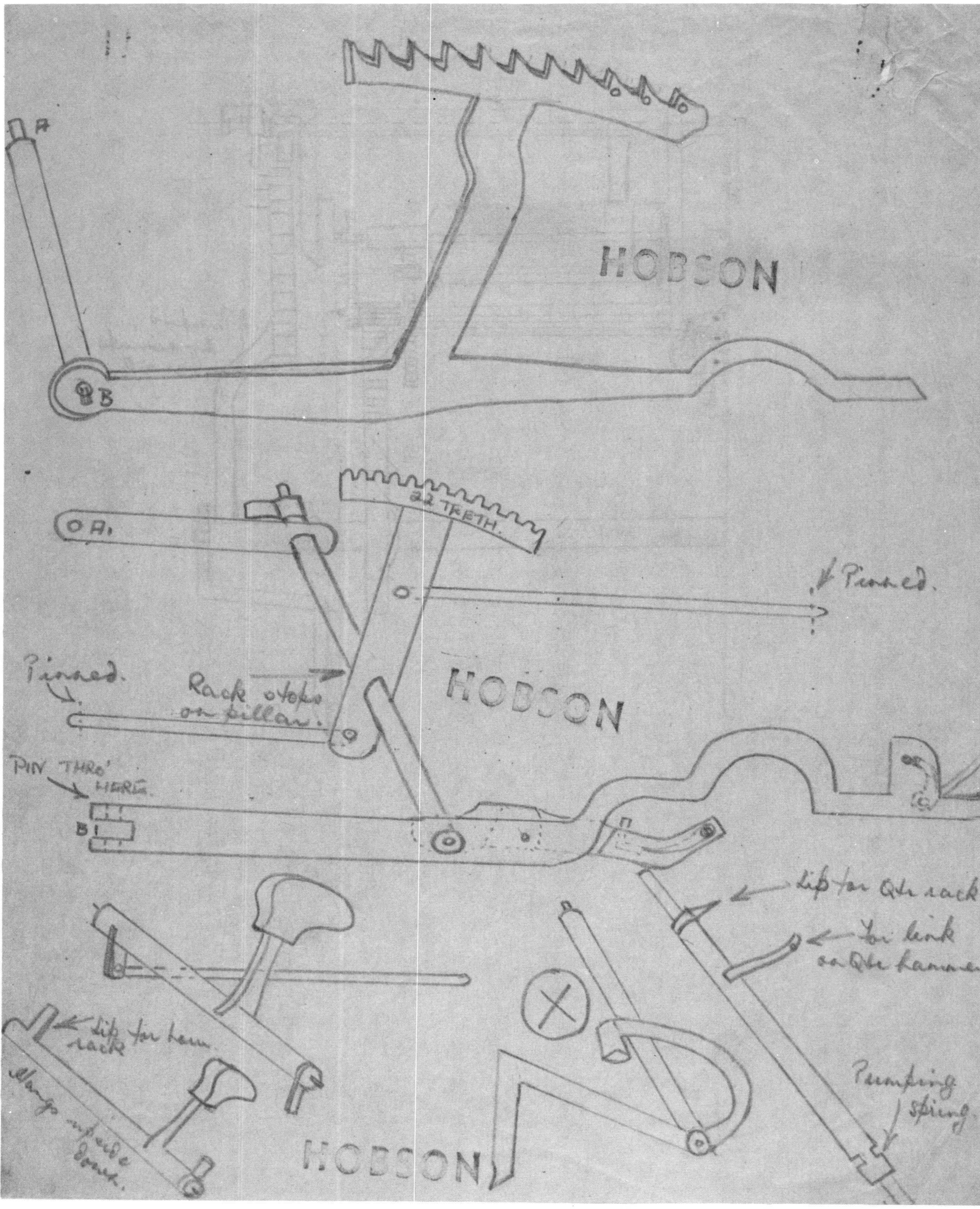

Tompion. Detail parts for use with pages 14 - 18.

Timepiece with Silent Pull sounding the Quarters on a Single Bell

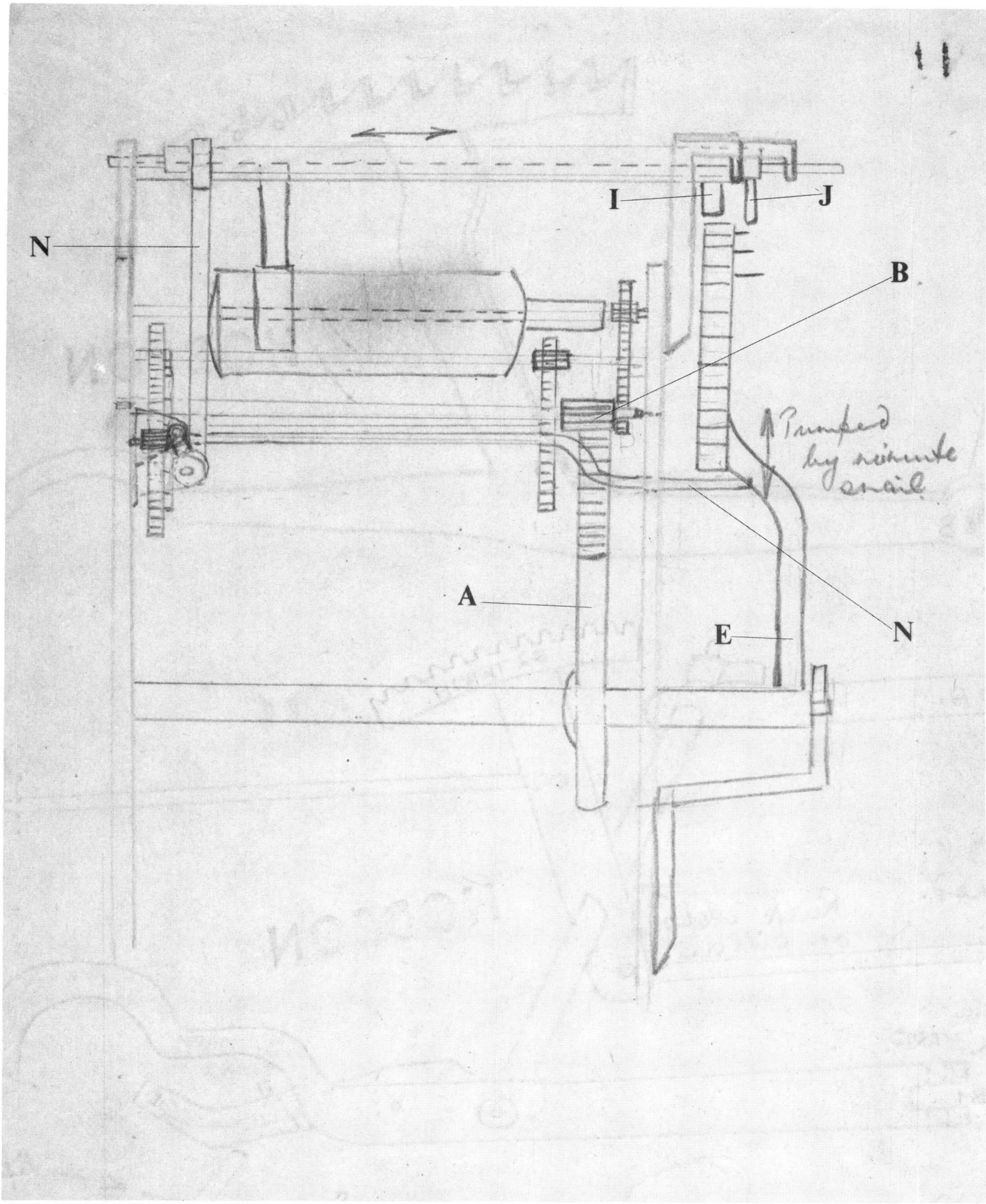

Tompion. Detail parts for use with pages 14 - 18, with particular reference to pages 12 and 13.

Timepiece with Silent Pull sounding the Quarters on a Single Bell

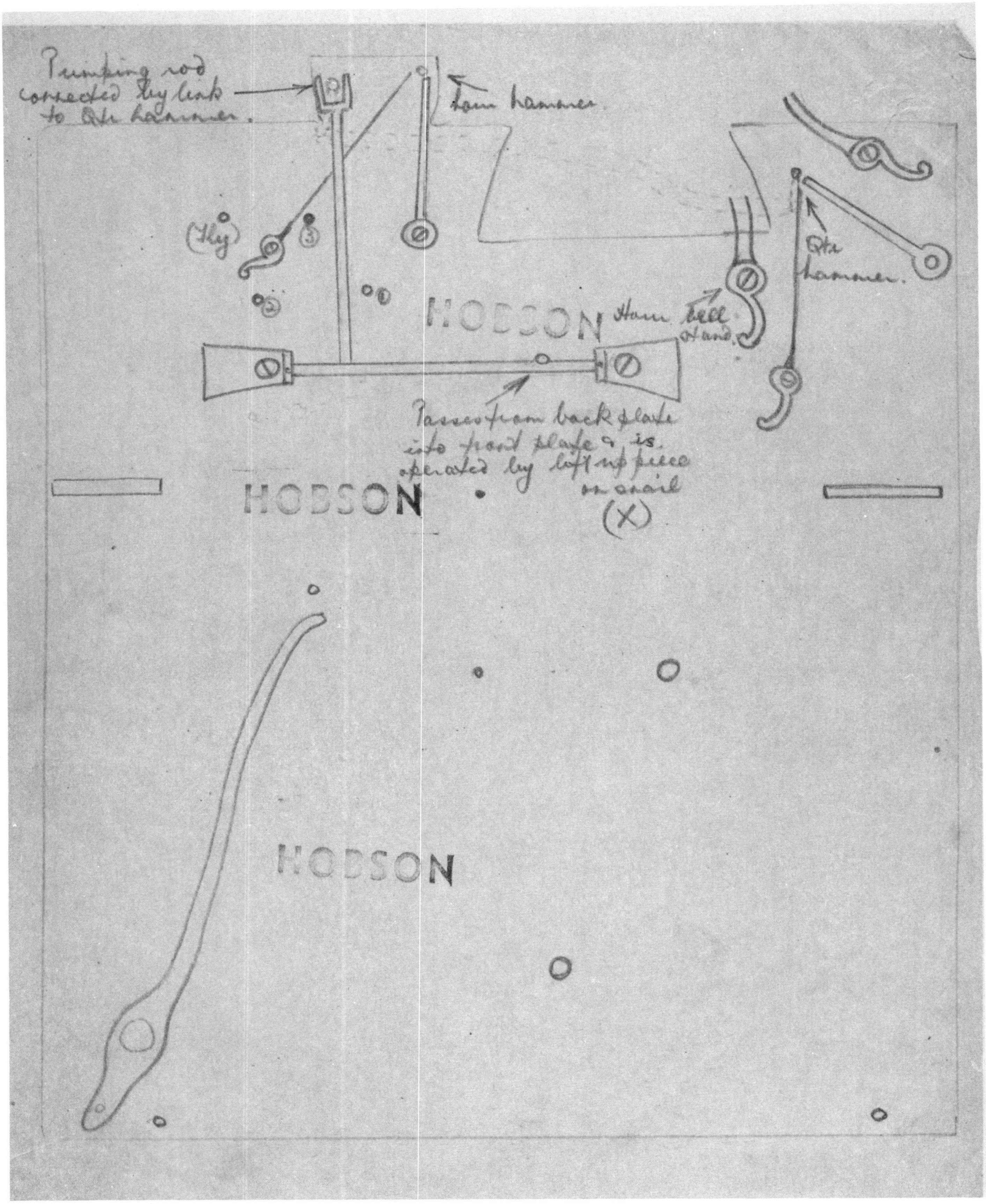

Tompion. View inside back plate showing pump lever, blade spring and slots for pull eyes. For use with pages 14 - 17.

Timepiece with Silent Pull sounding the Quarters on a Single Bell

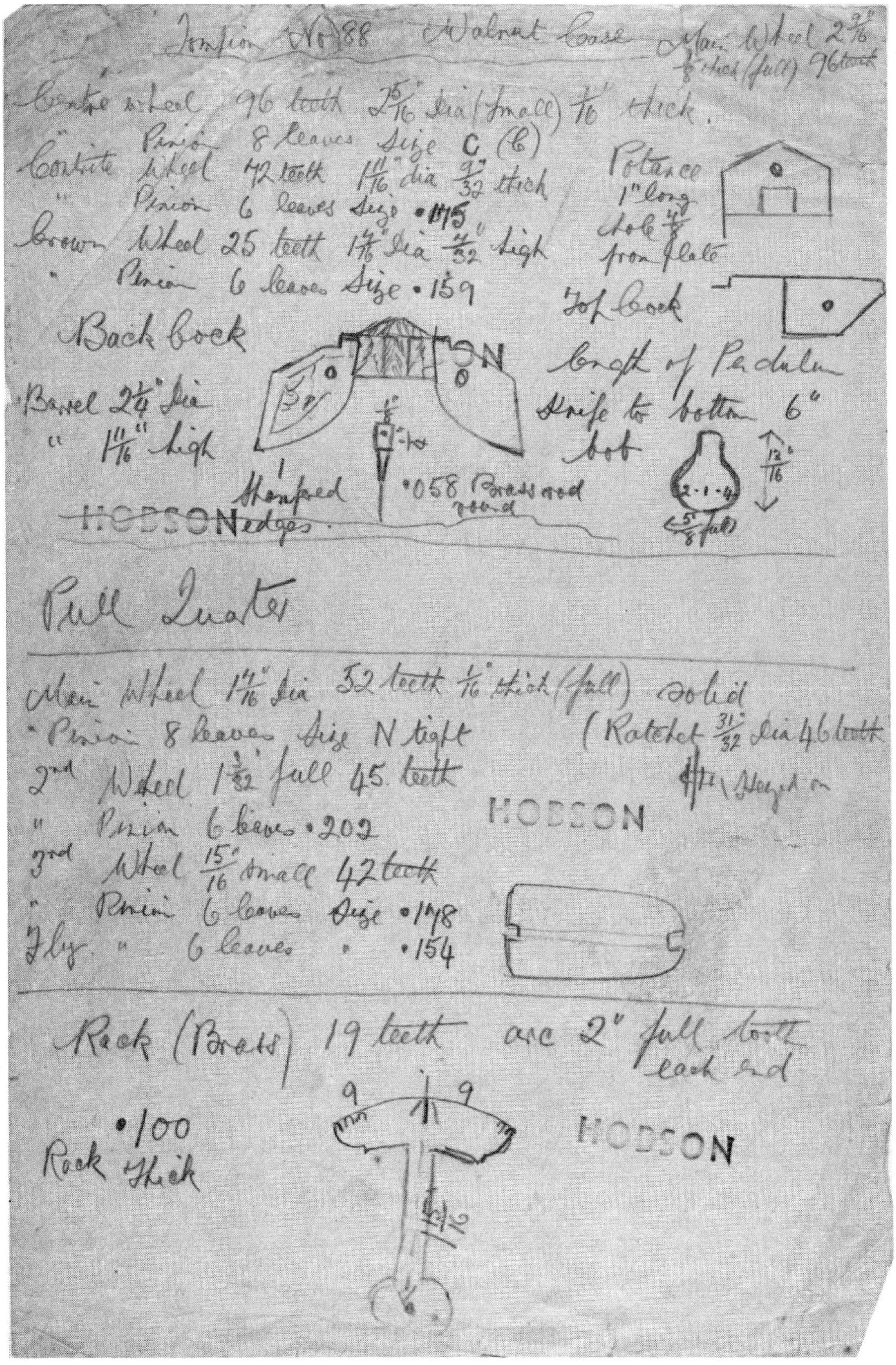

Tompion No 88 Walnut Case
Main Wheel 2 9/16" 1/8 thick (full) 96 teeth
Centre wheel 96 teeth 2 5/16" dia (small) 1/16" thick.
" Pinion 8 leaves Size C (B)
Contrite Wheel 72 teeth 1 11/16" dia 9/32" thick
" Pinion 6 leaves Size ·175
Crown Wheel 25 teeth 1 1/16" dia 4/32" high
" Pinion 6 leaves Size ·159
Potance 1" long Hole 4/8 from plate
Top Cock
Back Cock
Barrel 2 1/4" dia
" 1 11/16" high
Length of Pendulum Knife to bottom 6"
bob 13/16" 2·1·4 5/8 full
Stamped HOBSON edges. ·058 Brass rod round

Pull Quarter

Main Wheel 1 7/16" dia 32 teeth 1/16" thick (full) solid
Pinion 8 leaves Size N tight (Ratchet 31/32" dia 46 teeth
2nd Wheel 1 3/32 full 45 teeth Hinged on
" Pinion 6 leaves ·202
3rd Wheel 15/16" small 42 teeth
" Pinion 6 leaves Size ·178
Fly " 6 leaves " ·154
HOBSON

Rack (Brass) 19 teeth arc 2" full tooth each end
9 9
·100 Rack thick
HOBSON

Tompion No. 88. Details of wheel and pinion work. This seems to be the only known Tompion bracket clock in original walnut veneered case. See Symonds *Tompion*, fig. 119.

Timepiece with Silent Pull sounding the Quarters on a Single Bell

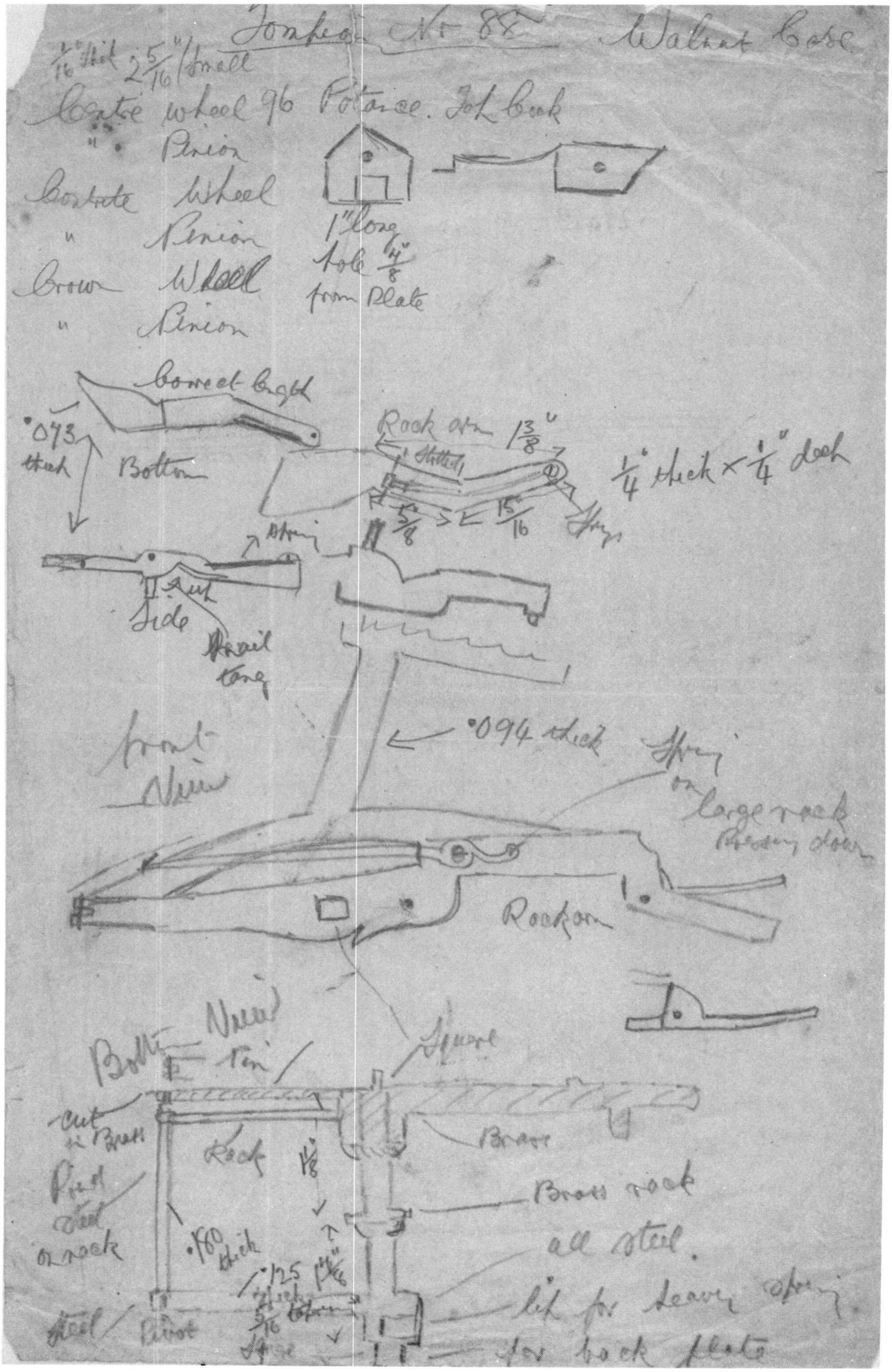

Tompion No. 88. Further details of this clock.

Timepiece with Silent Pull sounding the Quarters on a Single Bell

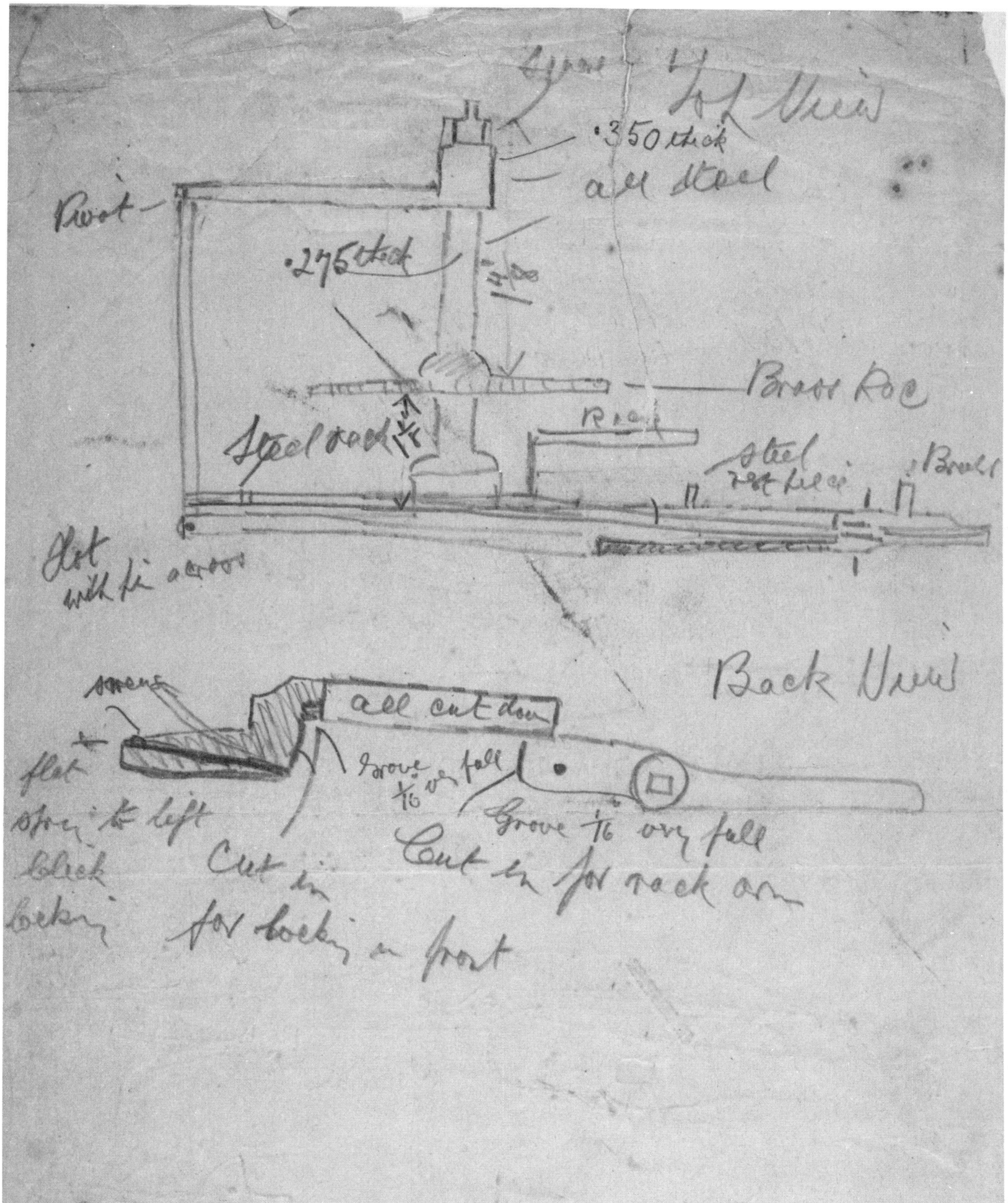

Tompion No. 88. Further details of this clock.

Tompion Silent Pull Sounding the Quarters on a Single Bell, as seen by Thiout in 1741

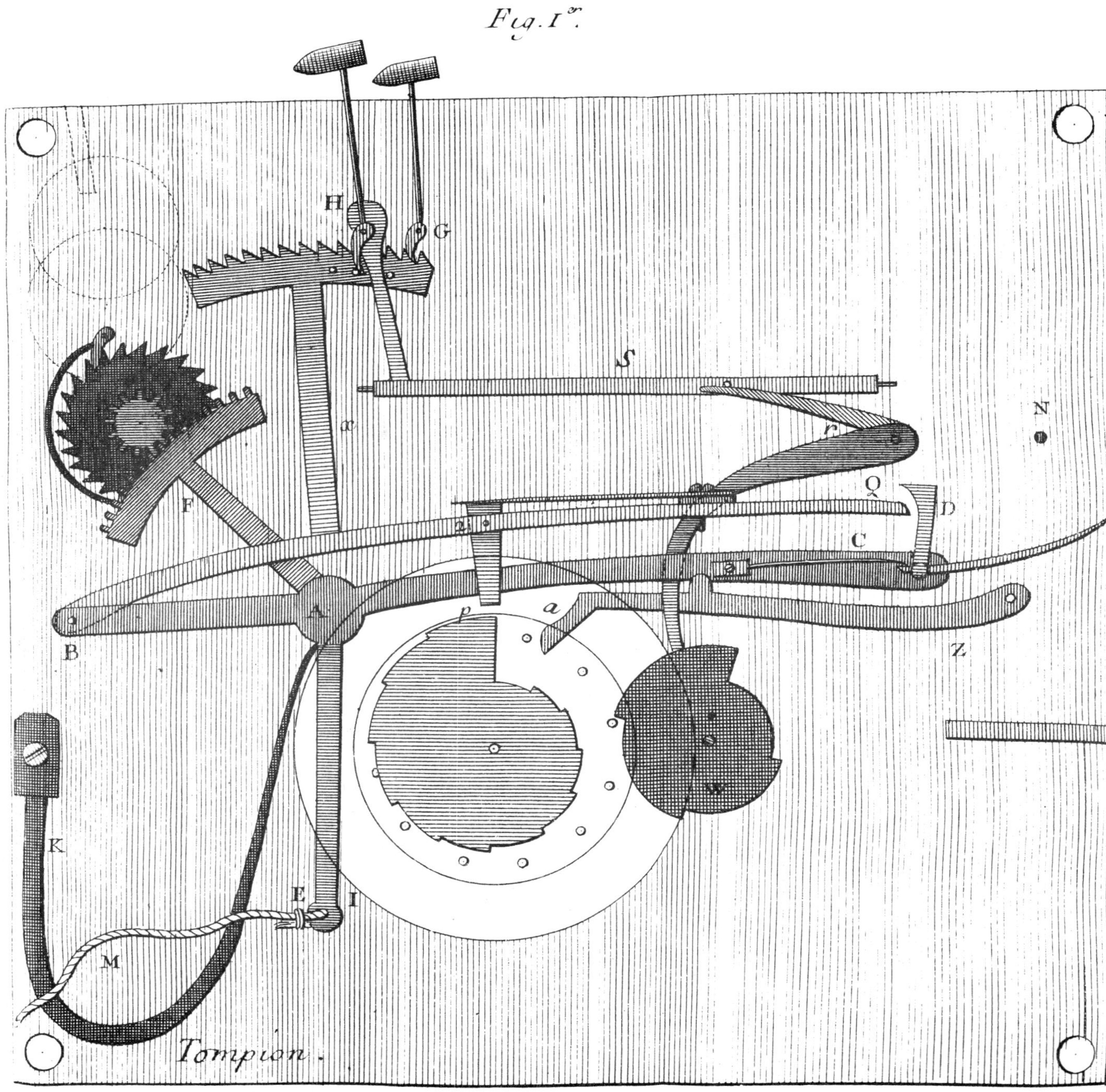

Tompion/Thiout. Thiout makes a remarkably good job of showing Tompion's standard form of silent pull despite a little artistic licence. Thiout's drawing is still worth studying today in conjunction with Hobson.

Striking Clock with Pull Quarter on a Single Bell

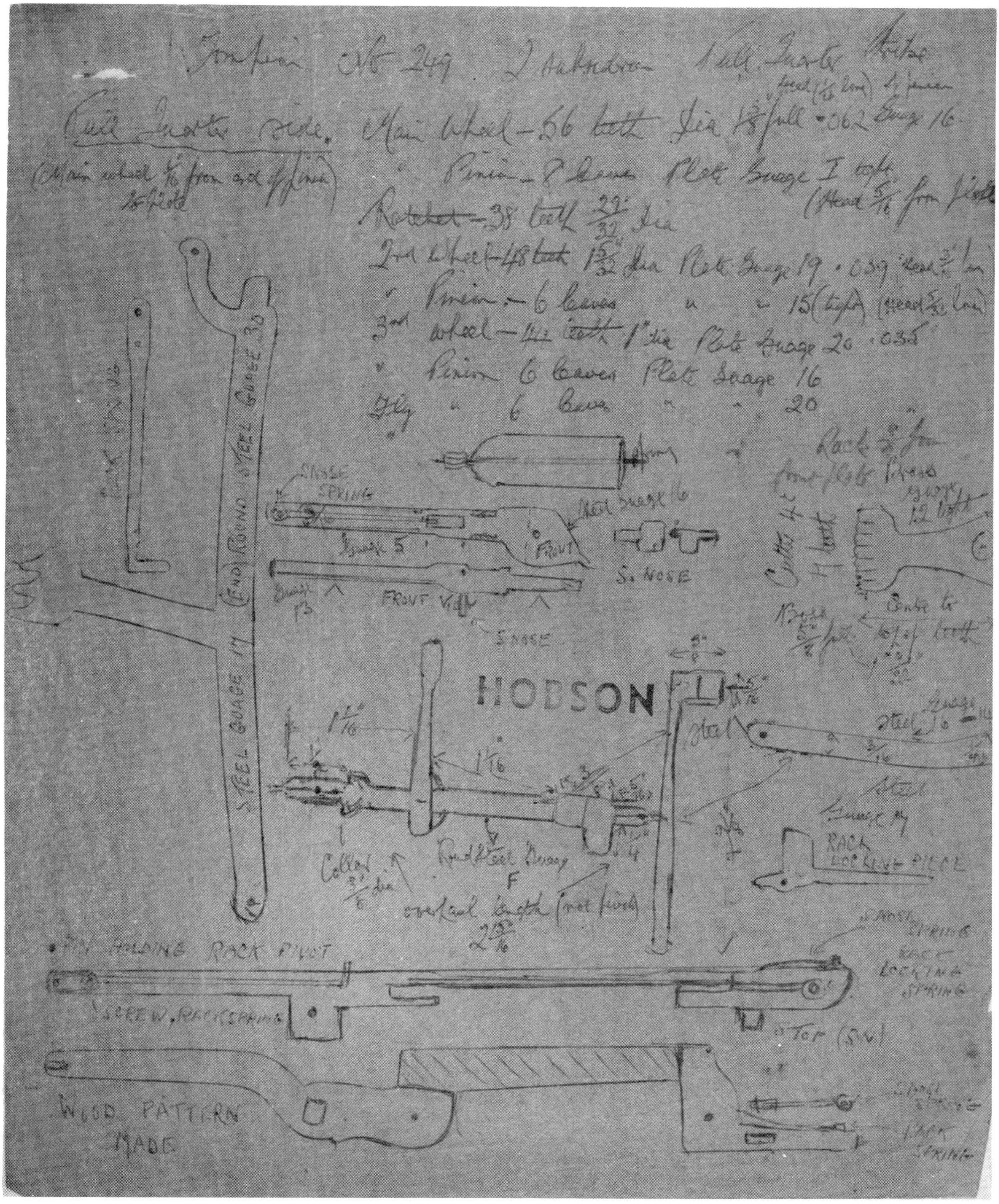

Tompion No. 249. Typical pull quarter parts as used in striking clocks throughout his career. The illustration on page 28 shows similar parts assembled under the dial of Graham No. 689.

Tompion's striking bracket clocks with pull quarter were made in three standard sizes. All employed an almost exactly similar movement layout.

Striking Clock with Pull Quarter on a Single Bell

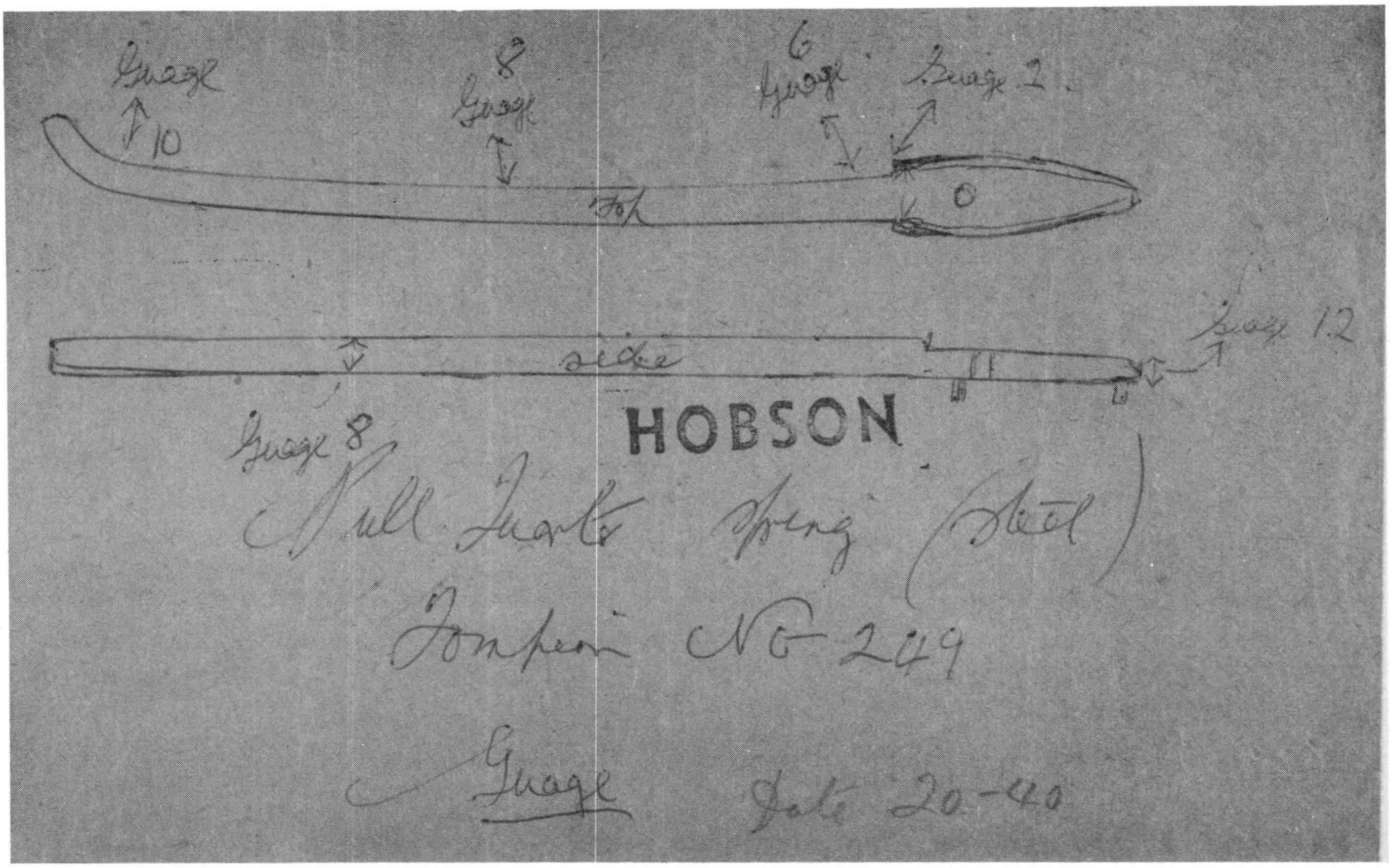

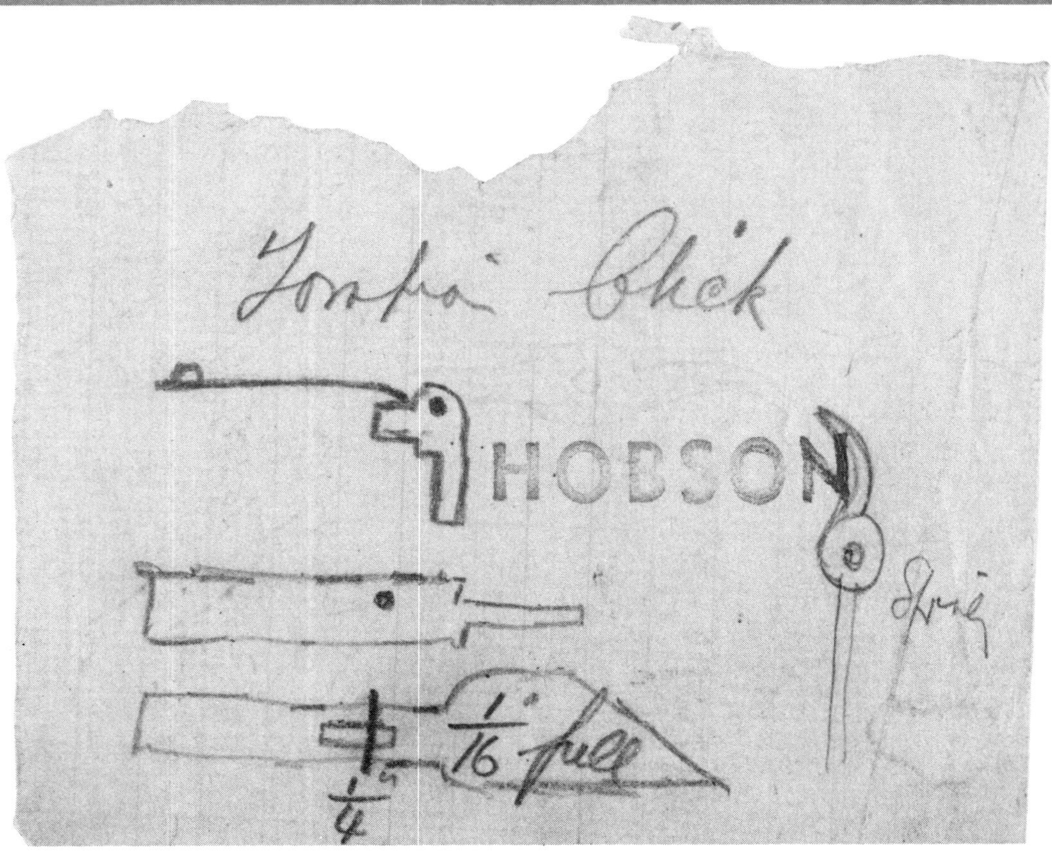

Tompion No. 249. Detail drawings of blade spring and pivoted quarter rack tail ("B" on page 28).

Striking Clock with Pull Quarter on a Single Bell

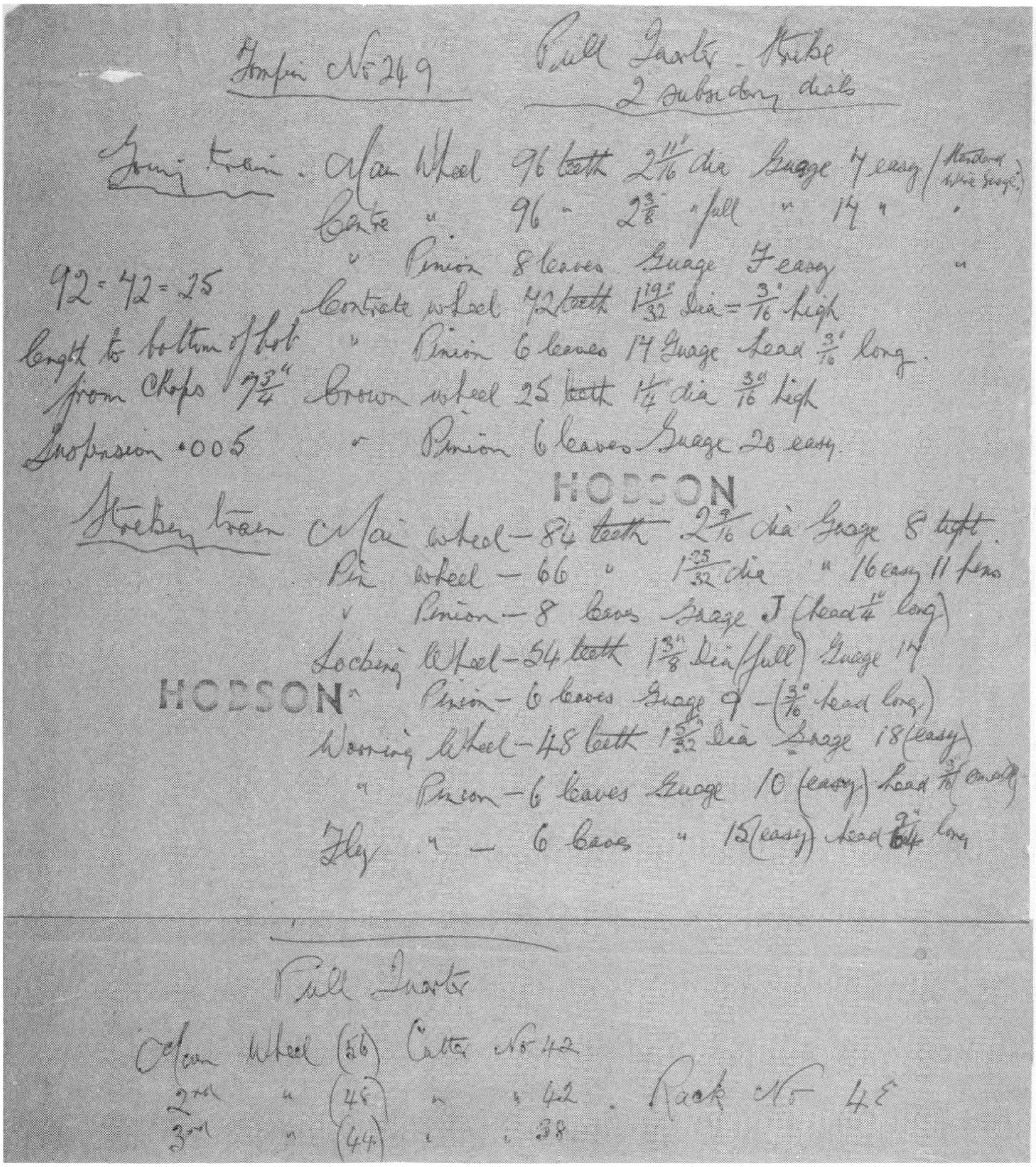

Tompion No 249 Pull Quarter – Strike
2 subsidiary dials

Going train. Main Wheel 96 teeth 2 11/16" dia Guage 7 easy (Standard Wire Guage)
Centre " 96 " 2 3/8 " full " 14 " "
" Pinion 8 leaves Guage 7 easy "
92 = 72 = 25
Contrate wheel 72 teeth 1 19/32 Dia = 3/16 high
Length to bottom of bob " Pinion 6 leaves 14 Guage head 3/16 long.
from Chops 7 3/4" Crown wheel 25 teeth 1 1/4 dia 3/16 high
Suspension ·005 " Pinion 6 leaves Guage 20 easy.

HOBSON

Striking train Main wheel – 84 teeth 2 9/16 dia Guage 8 tight.
Pin wheel – 66 " 1 25/32 dia " 16 easy 11 pins
" Pinion – 8 leaves Guage J (head 1/4" long)
Locking Wheel – 54 teeth 1 3/8" Dia (full) Guage 17
HOBSON " Pinion – 6 leaves Guage 9 – (3/16 head long)
Warning Wheel – 48 teeth 1 5/32 Dia Guage 18 (easy)
" Pinion – 6 leaves Guage 10 (easy) head 3/16 (small)
Fly " – 6 leaves " 15 (easy) head 9/64 long

Pull Quarter

Main Wheel (56) Cutter No 42
2nd " (48) " " 42 . Rack No 4E
3rd " (44) " " 38

Tompion. No. 249. Details of trains. Most of the cutter sizes shown throughout this book are Philip Thornton's numbers. His successors are still in business at the Old Bakehouse, Upper Tysoe, Leamington, Warwickshire.

Striking Clock with Pull Quarter on a Single Bell

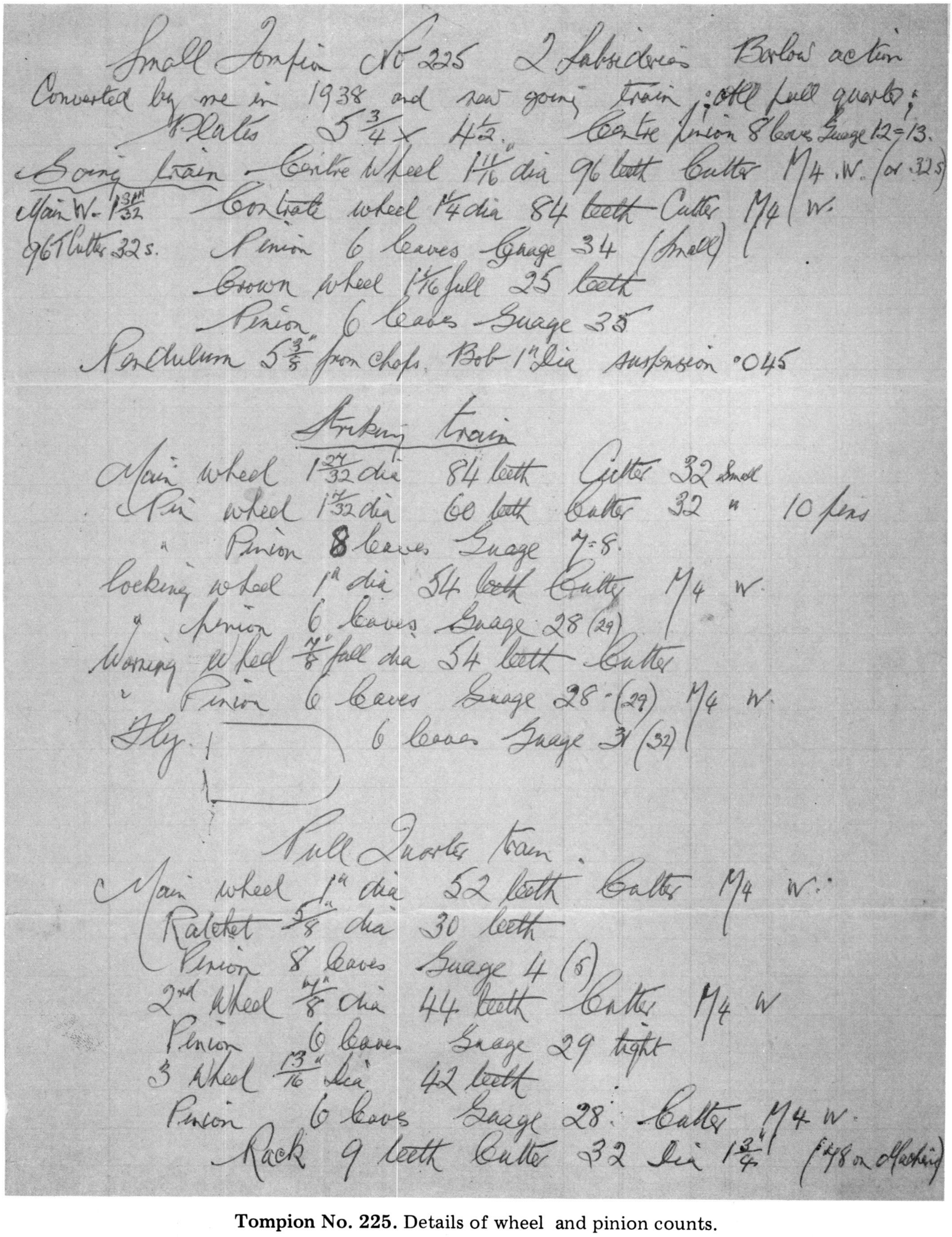
Small Tompion No 225 2 subsidiaries Barlow action
Converted by me in 1938 and new going train; old pull quarter;
Plates 5 3/4 x 4 1/2. Centre pinion 8 leaves Guage 12=13.
Going train — Centre wheel 1 11/16" dia 96 teeth Cutter M/4 W. (or 32s)
Main W. 1 31/32" Contrate wheel 1 1/4 dia 84 teeth Cutter M/4 W.
96 T Cutter 32s. Pinion 6 leaves Guage 34 (small)
Crown wheel 1 1/16 full 25 teeth
Pinion 6 leaves Guage 35
Pendulum 5 3/5" from chops. Bob 1" dia suspension ·045

Striking train
Main wheel 1 27/32 dia 84 teeth Cutter 32 small
Pin wheel 1 1/32 dia 60 teeth Cutter 32 " 10 pins
" Pinion 8 leaves Guage 7=8.
locking wheel 1" dia 54 teeth Cutter M/4 W.
" pinion 6 leaves Guage 28 (29)
Warning Wheel 7/8" full dia 54 teeth Cutter
" Pinion 6 leaves Guage 28 · (29) M/4 W.
Fly. 6 leaves Guage 31 (32)

Pull Quarter train
Main wheel 1" dia 52 teeth Cutter M/4 W.
Ratchet 5/8" dia 30 teeth
Pinion 8 leaves Guage 4 (5)
2nd Wheel 6/8" dia 44 teeth Cutter M/4 W
Pinion 6 leaves Guage 29 tight
3 Wheel 13/16" dia 42 teeth
Pinion 6 leaves Guage 28. Cutter M/4 W.
Rack 9 teeth Cutter 32 dia 1 3/4" (48 on Marking)

Tompion No. 225. Details of wheel and pinion counts.

Striking Clock with Pull Quarter on a Single Bell

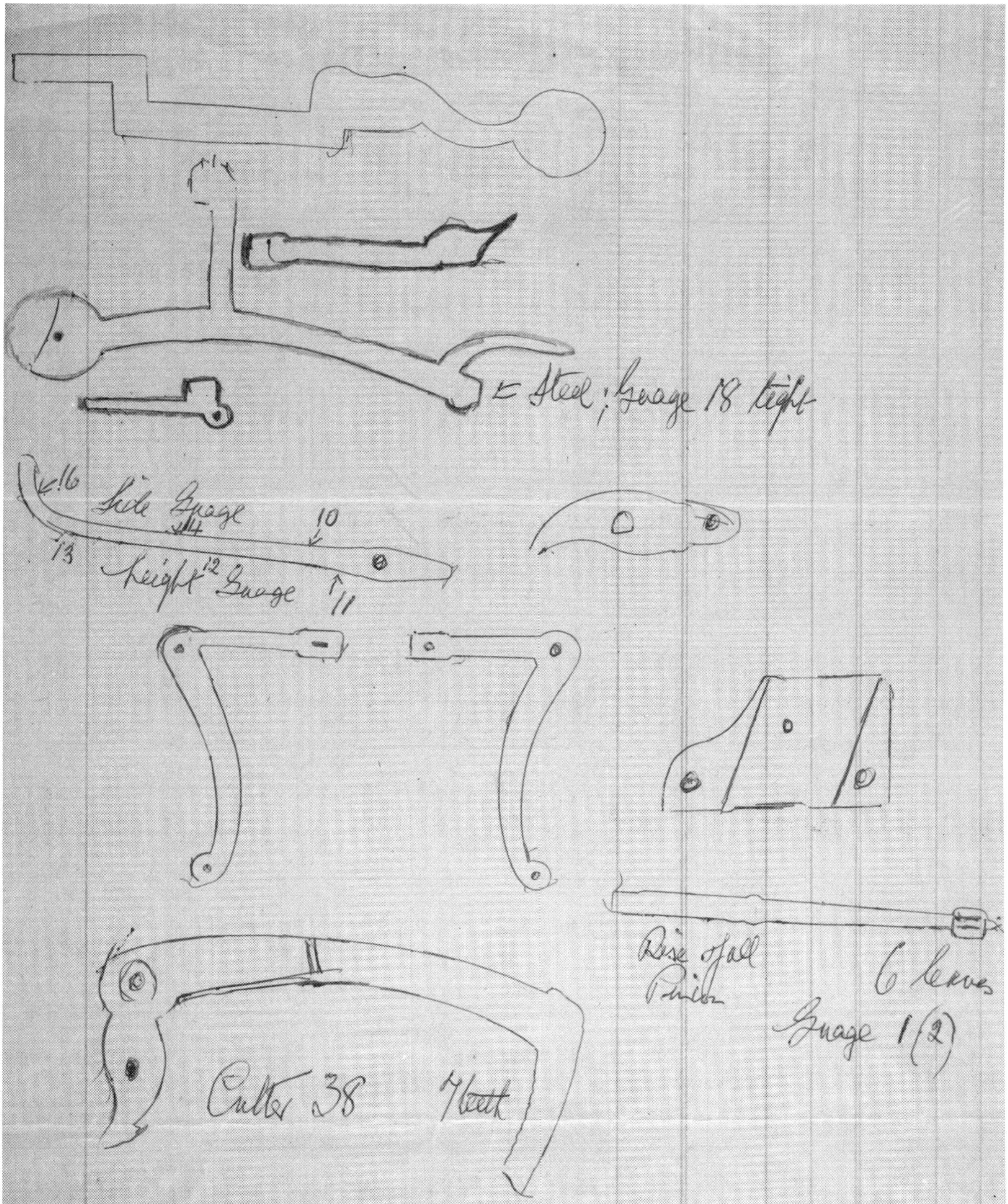

Tompion No. 225. Details of various parts.

GRAHAM
No. 689

Striking Clock with Pull Quarter on a Single Bell

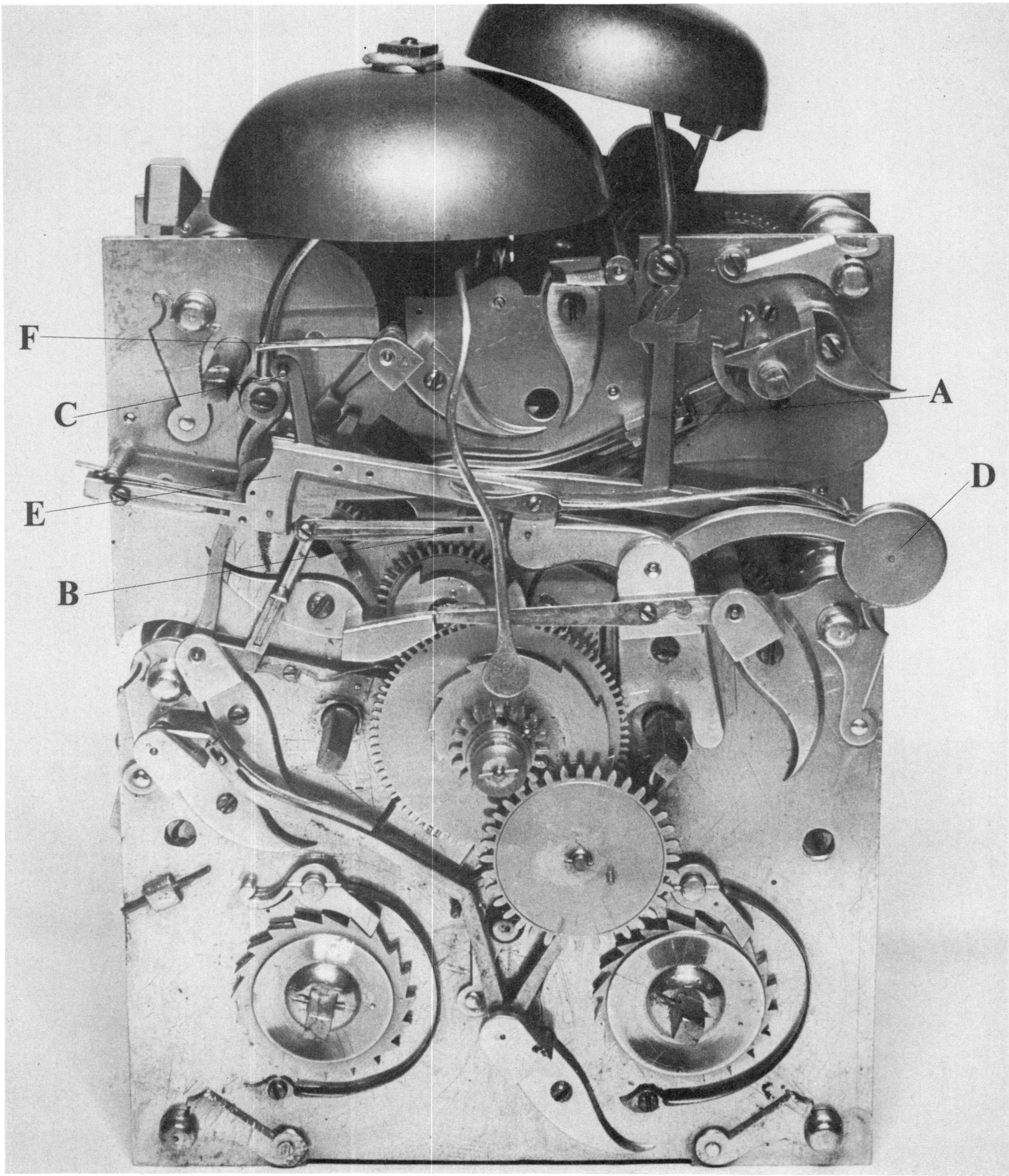

Graham. Tompion's pull quarter. On pulling either of the cords (see page 35 right) the internal motion rack turns anti-clockwise gathering the clickwork as already described on page 13 in connection with silent pull. In this striking clock the large quarter rack assembly ADE is pivoted on the right side. The quarter rack part A which has three teeth is raised at the end of the pull and lowered at the end of the run as in a silent pull. Note that in pull quarter clocks the pivoted rack tail B works with the quarter snail instead of with the hour snail. The hour striking is warned and released on the completion of the quarters by the pin C carried on the upright arm of the quarter rack. This pin lifts the arm F which in turn raises the internal hour rack hook.

Striking Clock with Pull Quarter on a Single Bell

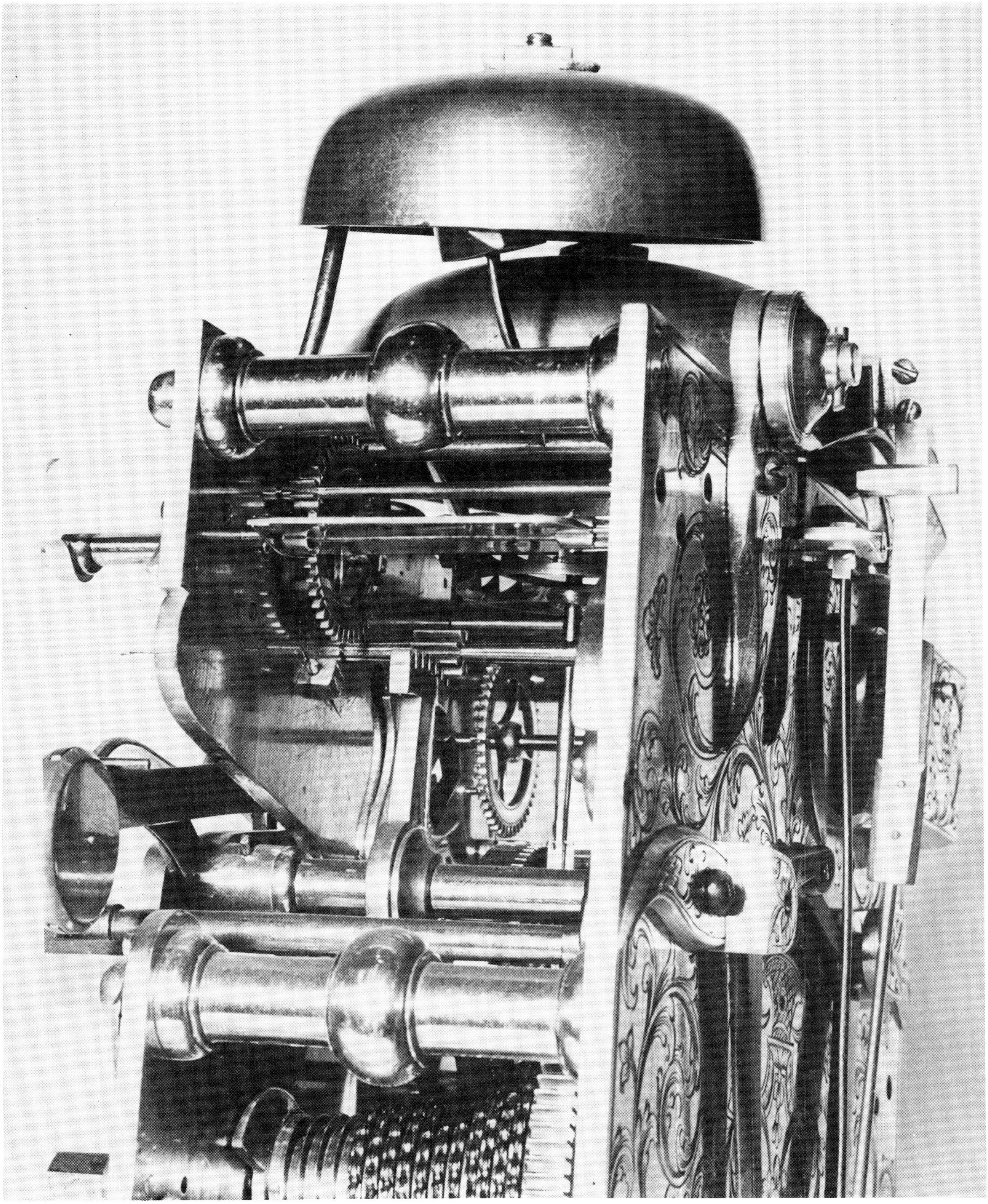

Graham. Tompion's pull quarter. In this side view note the internal motion rack and pinion. Opposite (bottom) note the most ingenious "V"-shaped pivoted lever which provides maintaining by means of a bolt acting on the centre wheel when the pipe of the winding key displaces a crescent-shaped shroud on the right of the square.

TOMPION & BANGER
No. 457

Striking Clock with Pull Quarter on a Single Bell

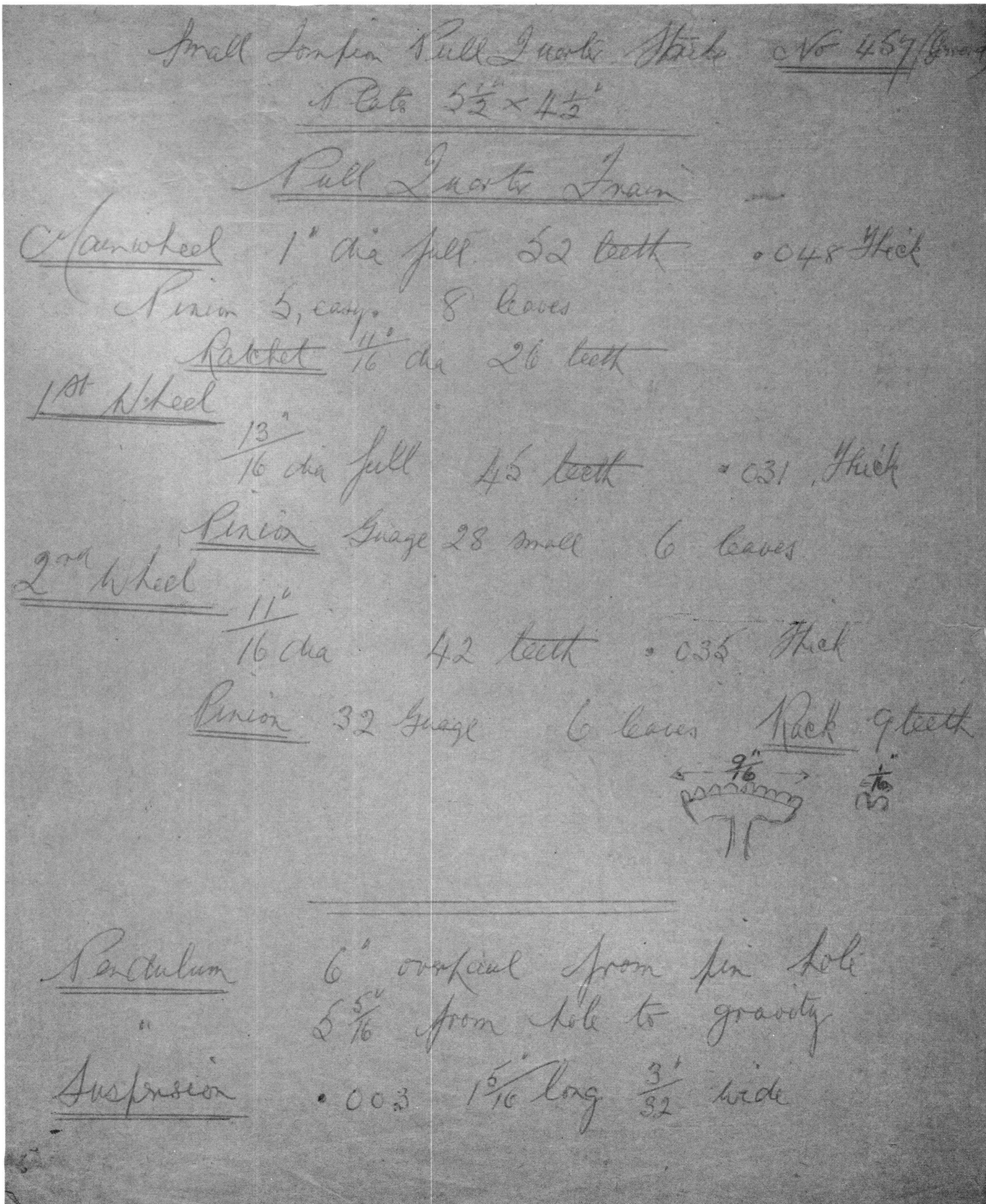
Small Tompion Pull Quarter Strike No 457 (Banger)
Plate 5½" × 4½"

Pull Quarter Train

Mainwheel 1" dia full 52 teeth .048 Thick
Pinion 5, easy. 8 leaves
Ratchet 11/16" dia 26 teeth
1st Wheel
13/16" dia full 45 teeth .031 Thick
Pinion Guage 28 small 6 leaves
2nd Wheel
11/16" dia 42 teeth .035 Thick
Pinion 32 Guage 6 leaves Rack 9 teeth
9/16" 1/16"

Pendulum 6" overall from pin hole
" 5 5/16" from hole to gravity
Suspension .003 1 5/16 long 3/32" wide

Tompion & Banger No. 457. Details of wheel counts of a miniature clock. See Symonds *Tompion*, **fig. 198.**

Striking Clock with Pull Quarter on a Single Bell

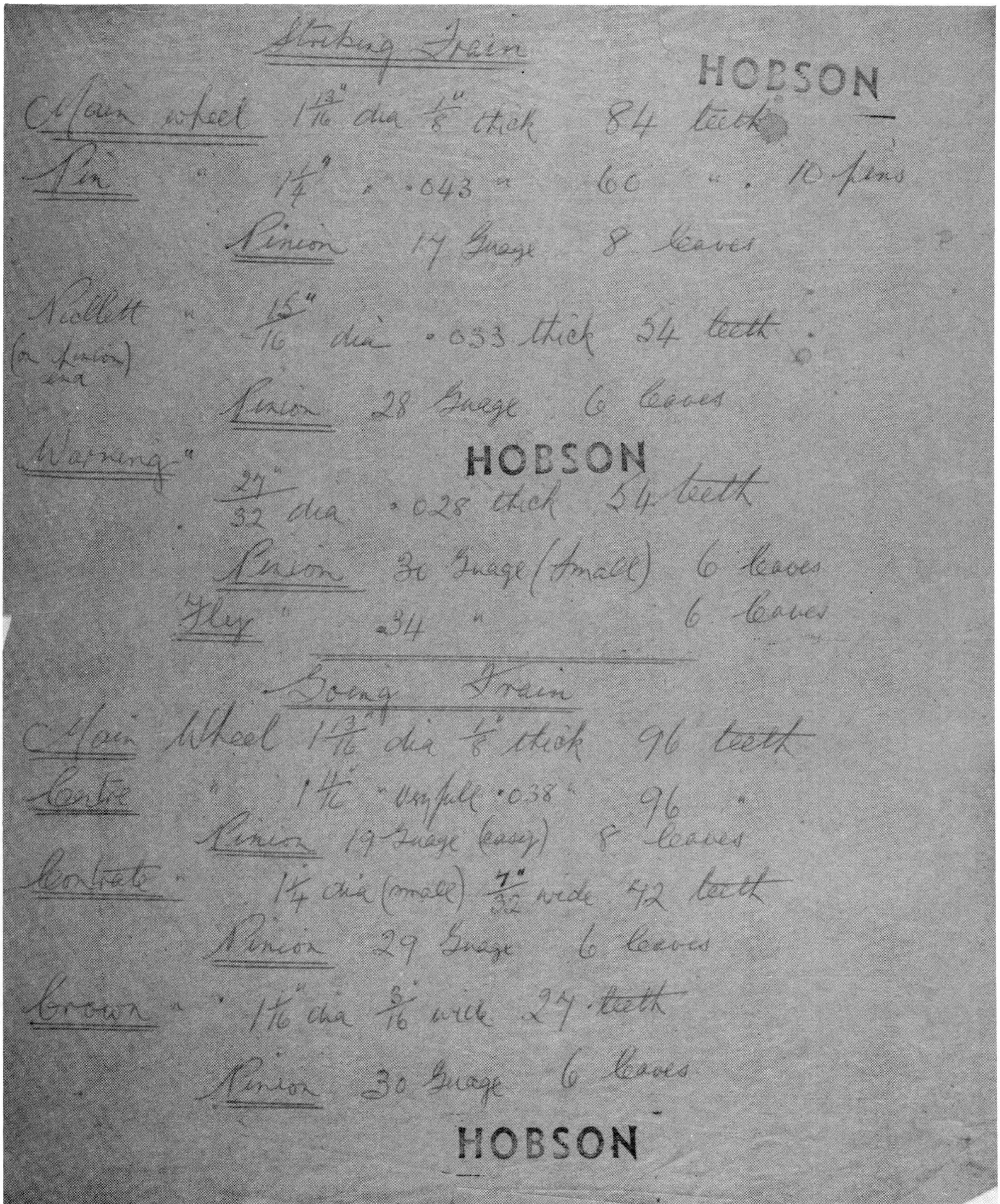

Tompion & Banger No. 457. Further details of work.

Striking Clock with Pull Quarter on a Single Bell

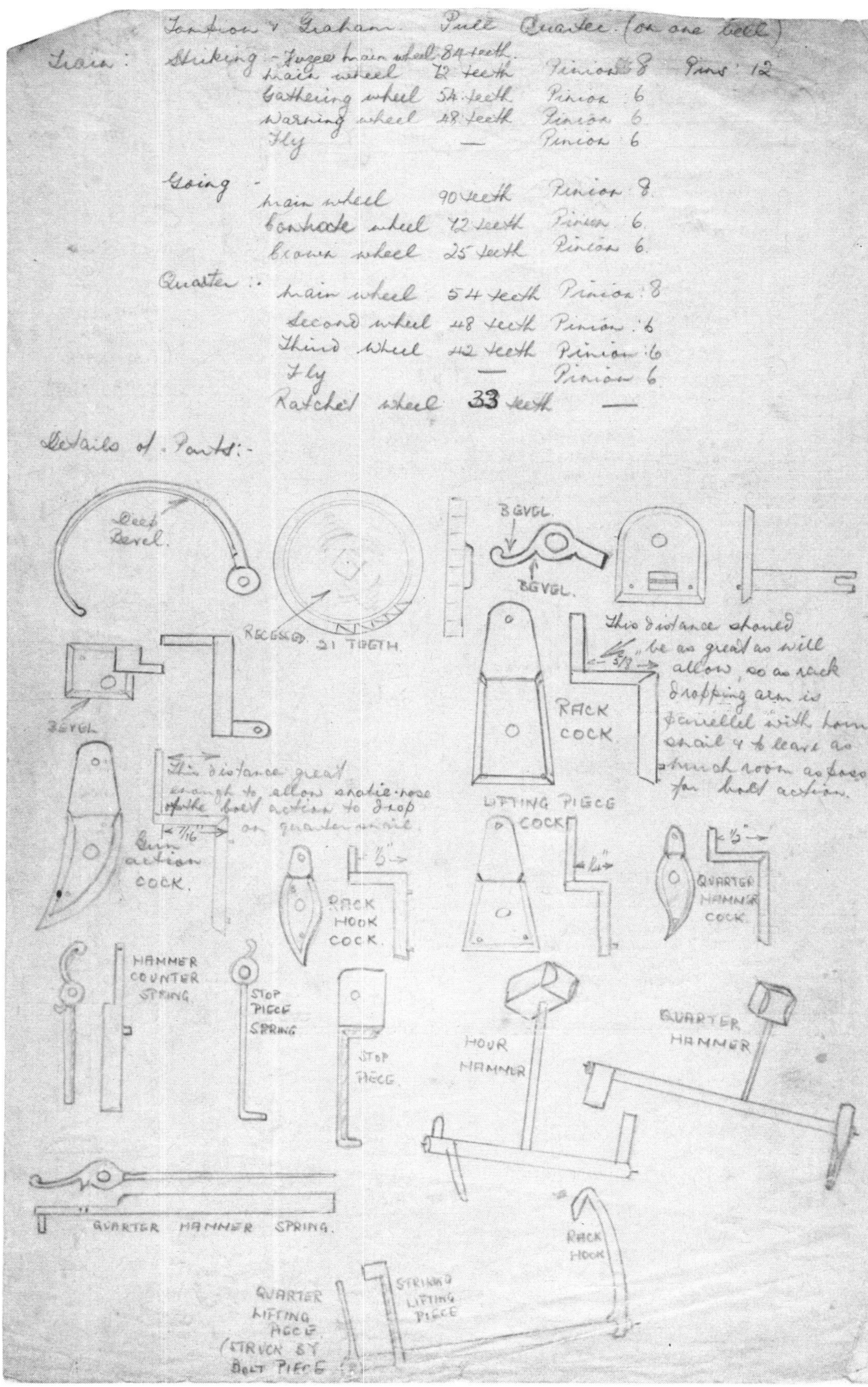

Tompion & Graham. Wheel counts of trains and details of certain small parts. The number of the clock is not recorded but the piece is virtually identical to that shown on page 28.

Striking Clock with Pull Quarter on a Single Bell

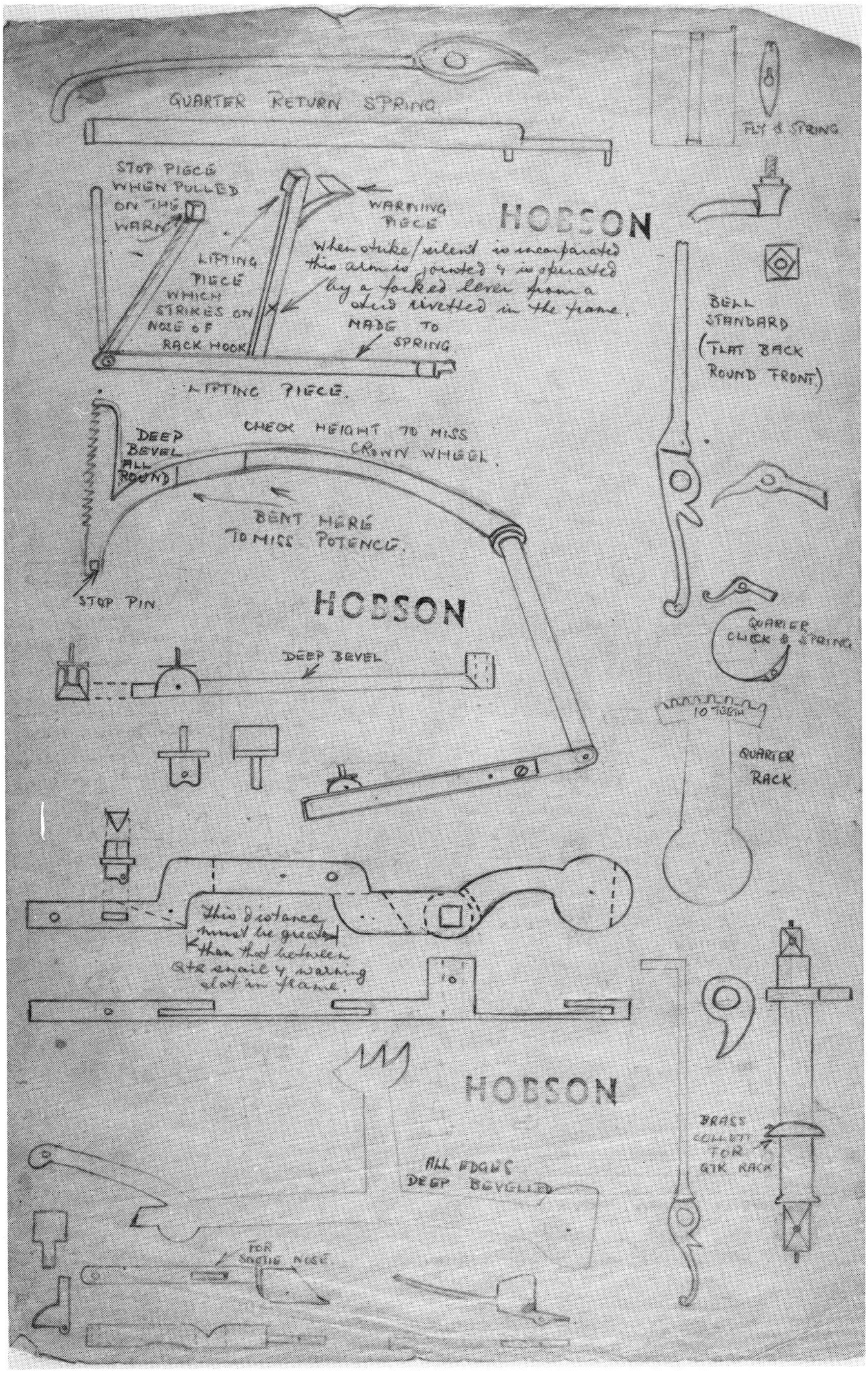

Tompion & Graham. Further details of parts of clock shown on page 32.

Striking Clock with Pull Quarter on a Single Bell

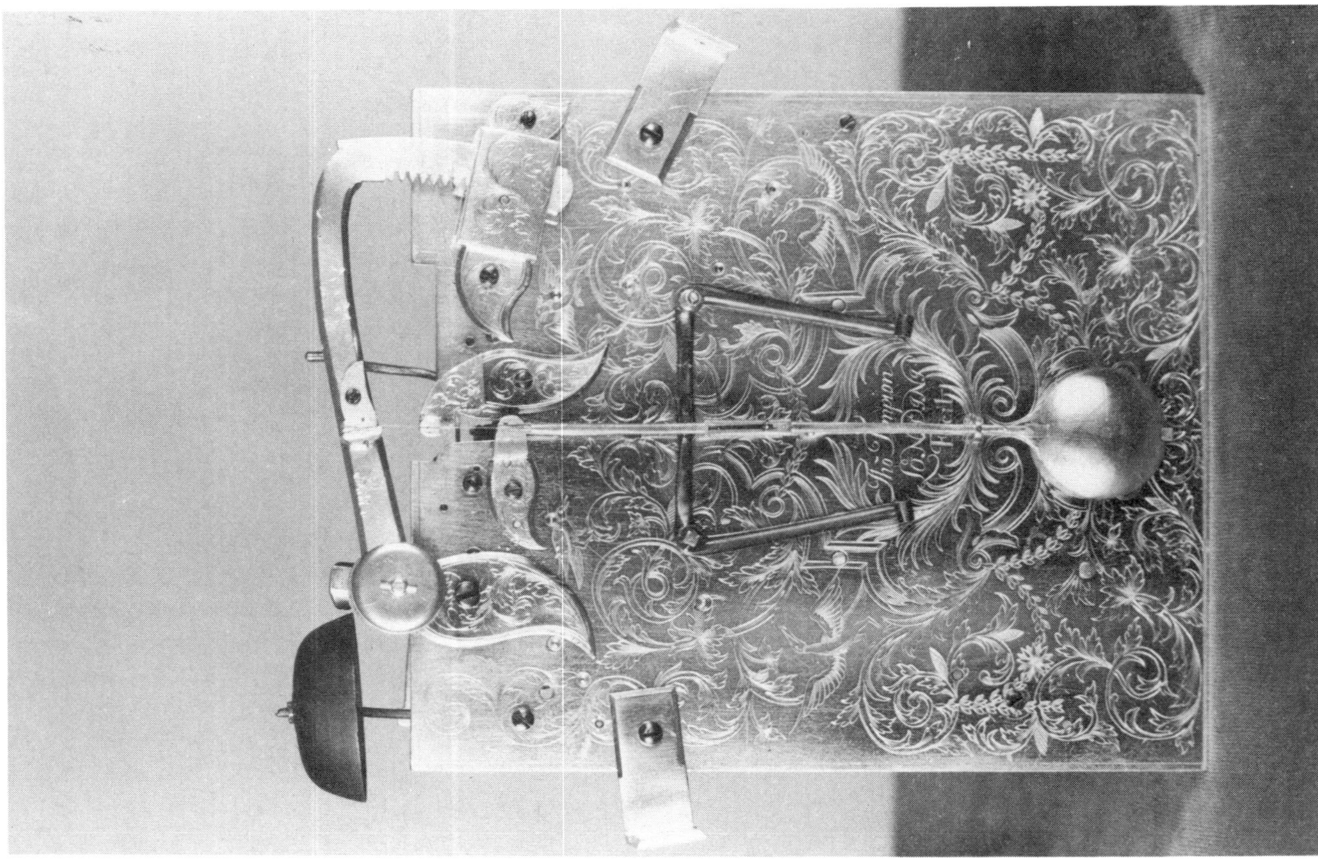

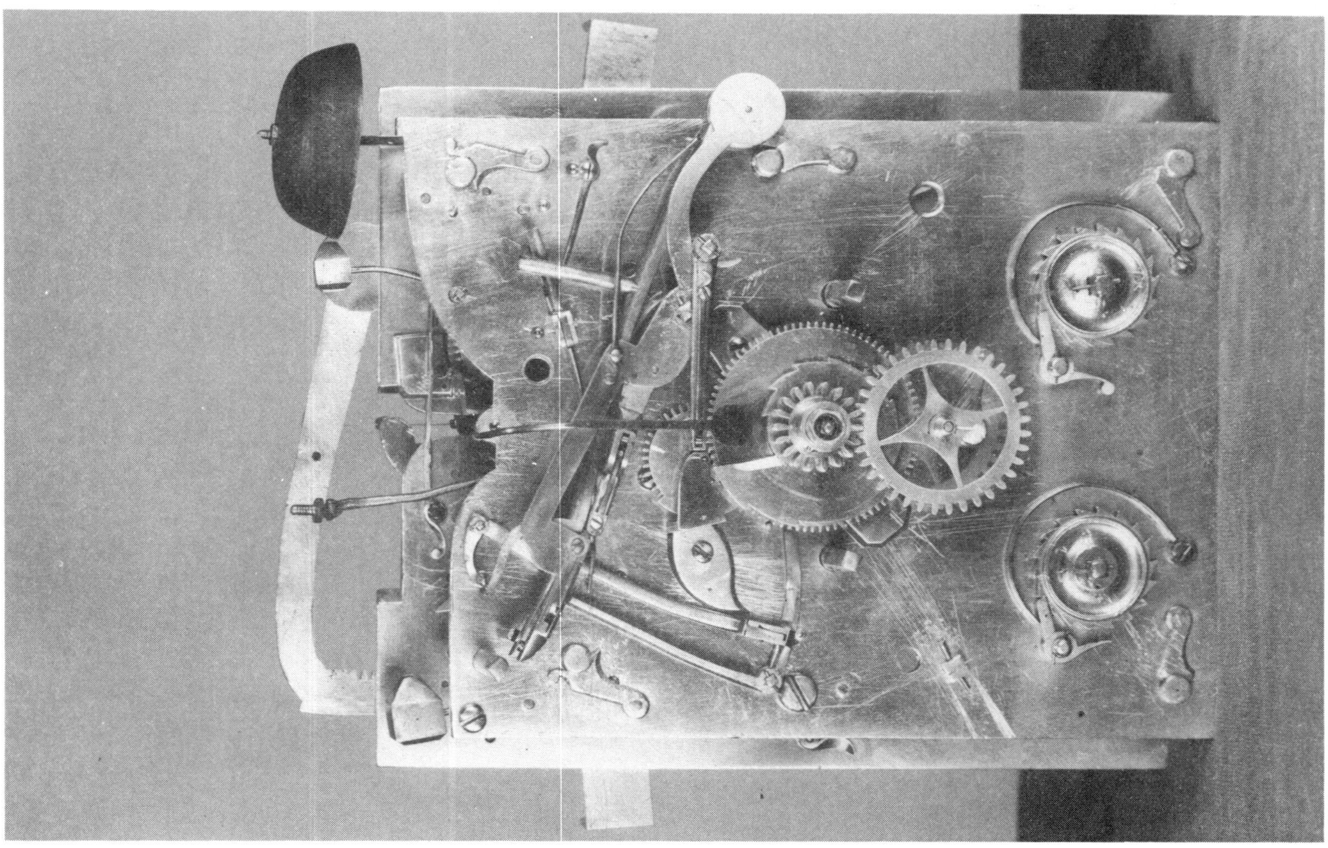

Tompion. A pull quarter clock with un-cut front plate.

Pull Arms as used in early Timepiece and late Striking Clocks

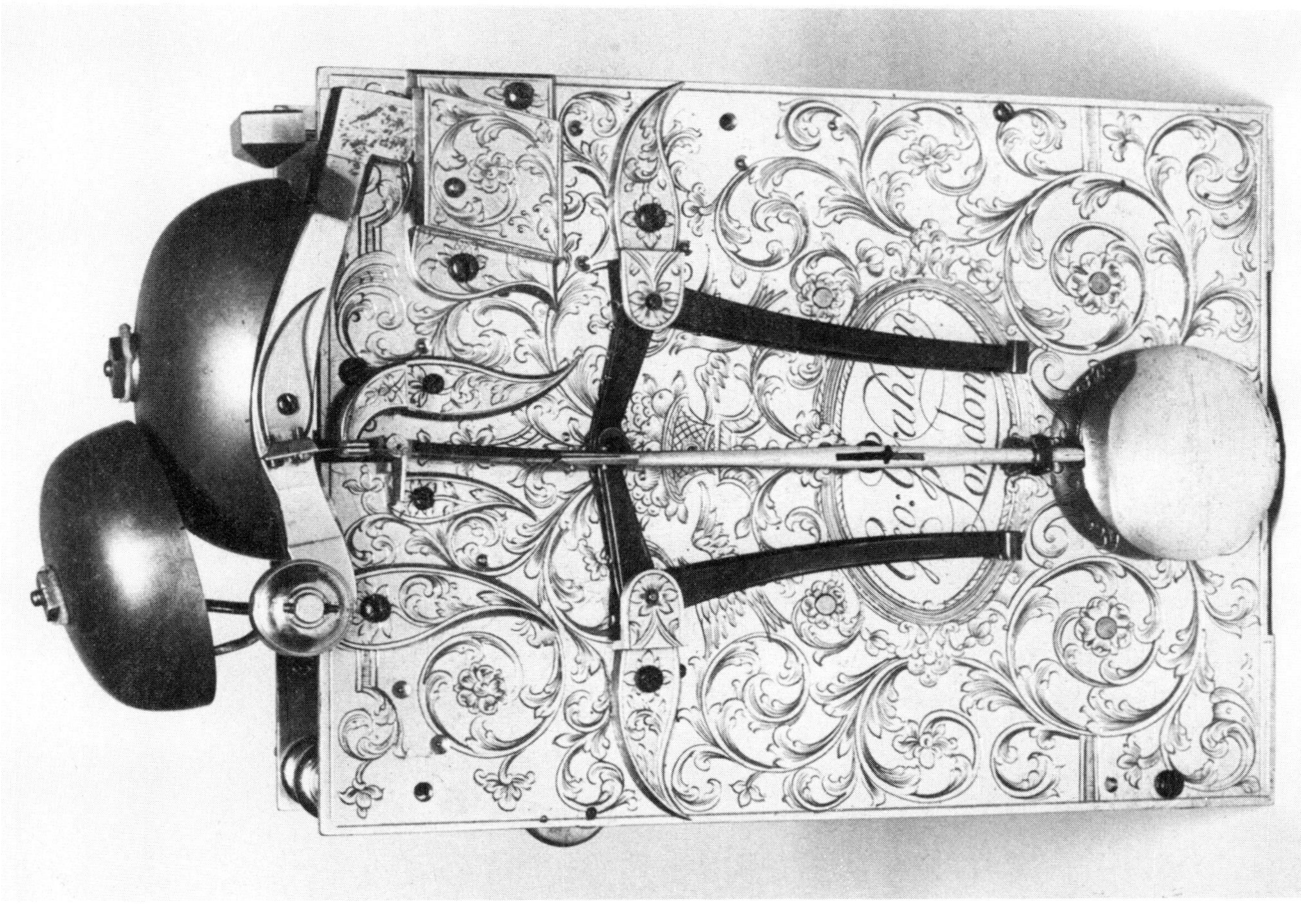

Graham No. 689. Late pull quarter. Here the pull is required only to cock the blade spring sufficiently to sound the quarters.

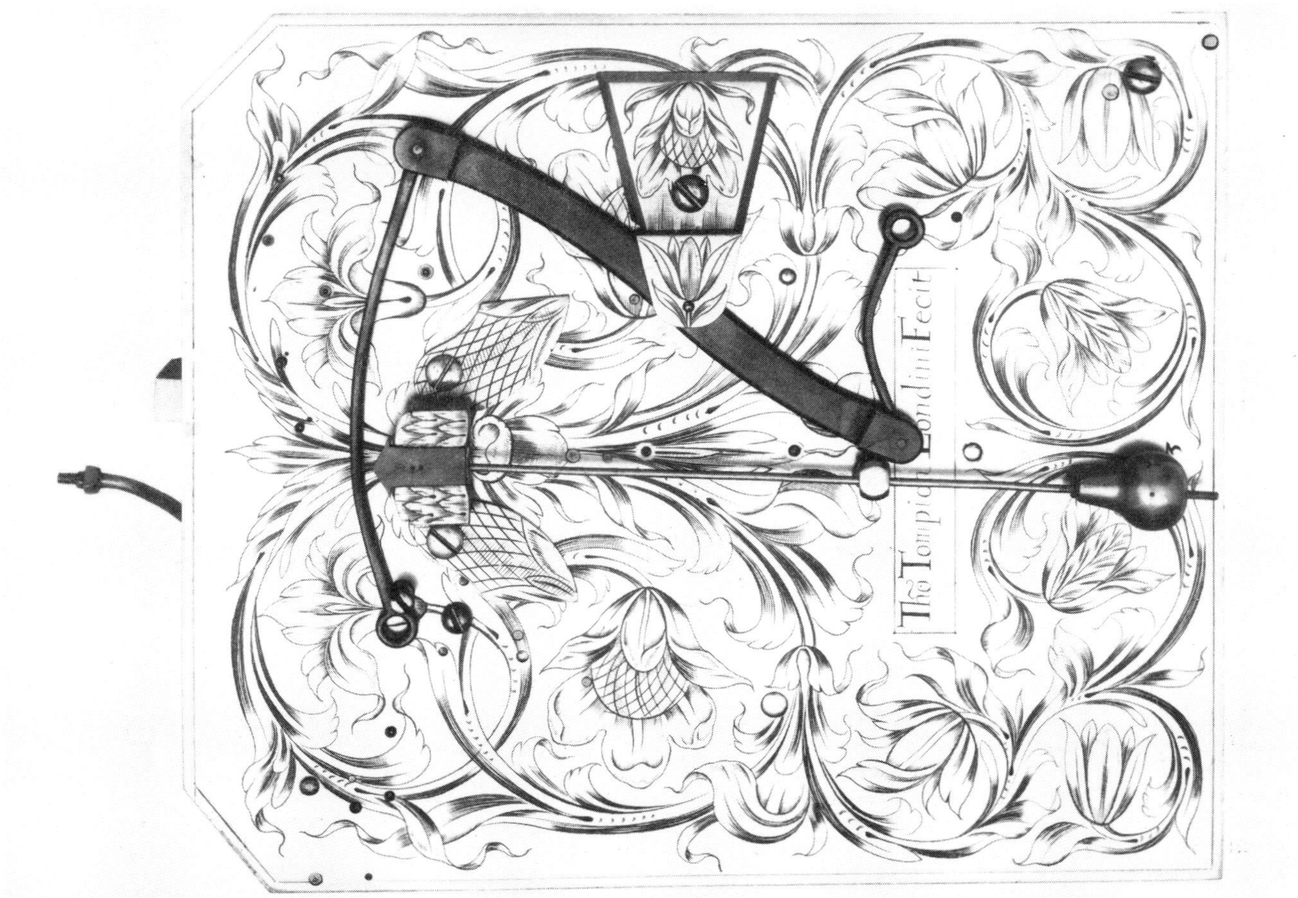

Tompion No. 93. Early silent pull. Note a system allowing the long pull required to cock a blade spring sufficiently to sound both hours and quarters.

Timepiece with Silent Pull sounding the Quarters on a Single Bell and with Passing Strike

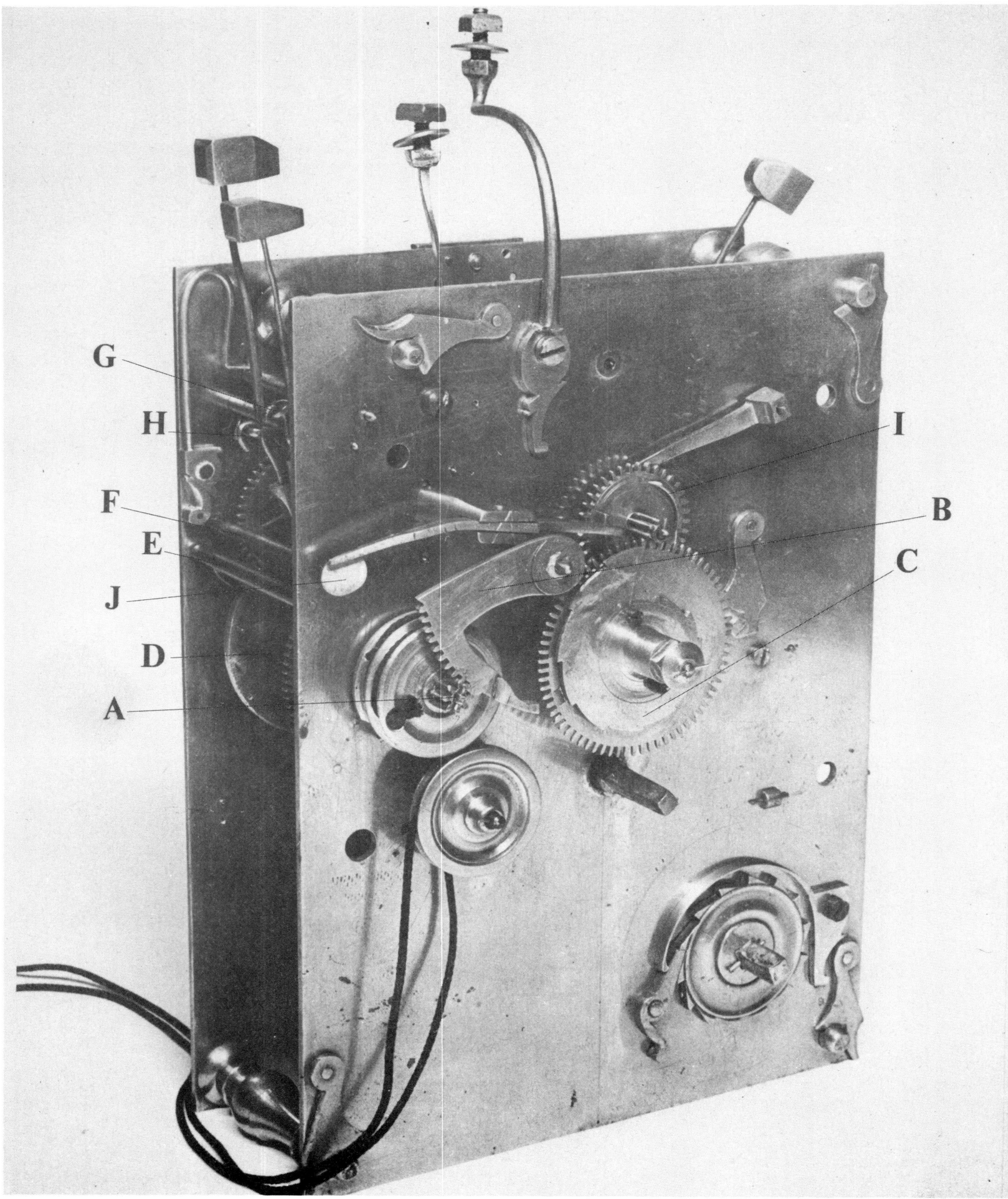

Joseph Knibb. Typical silent pull as used from c.1680 by the Knibb brothers. Knibb's system is exceedingly simple and straightforward. It employs the minimum number of parts necessary to achieve the result. In this respect it is the exact opposite of Tompion's. The Knibb-type silent pull, unlike that of Tompion, served as the inspiration for many other makers. Tompion's system, so far as is known, was never copied other than by his successor and by some of his apprentices. Incidentally, this particular Knibb timepiece strikes one blow at each hour in passing by means of a pin on the minute wheel. The feature is extremely rare in clocks of this kind.

Timepiece with Silent Pull sounding the Quarters on a Single Bell and with Passing Strike

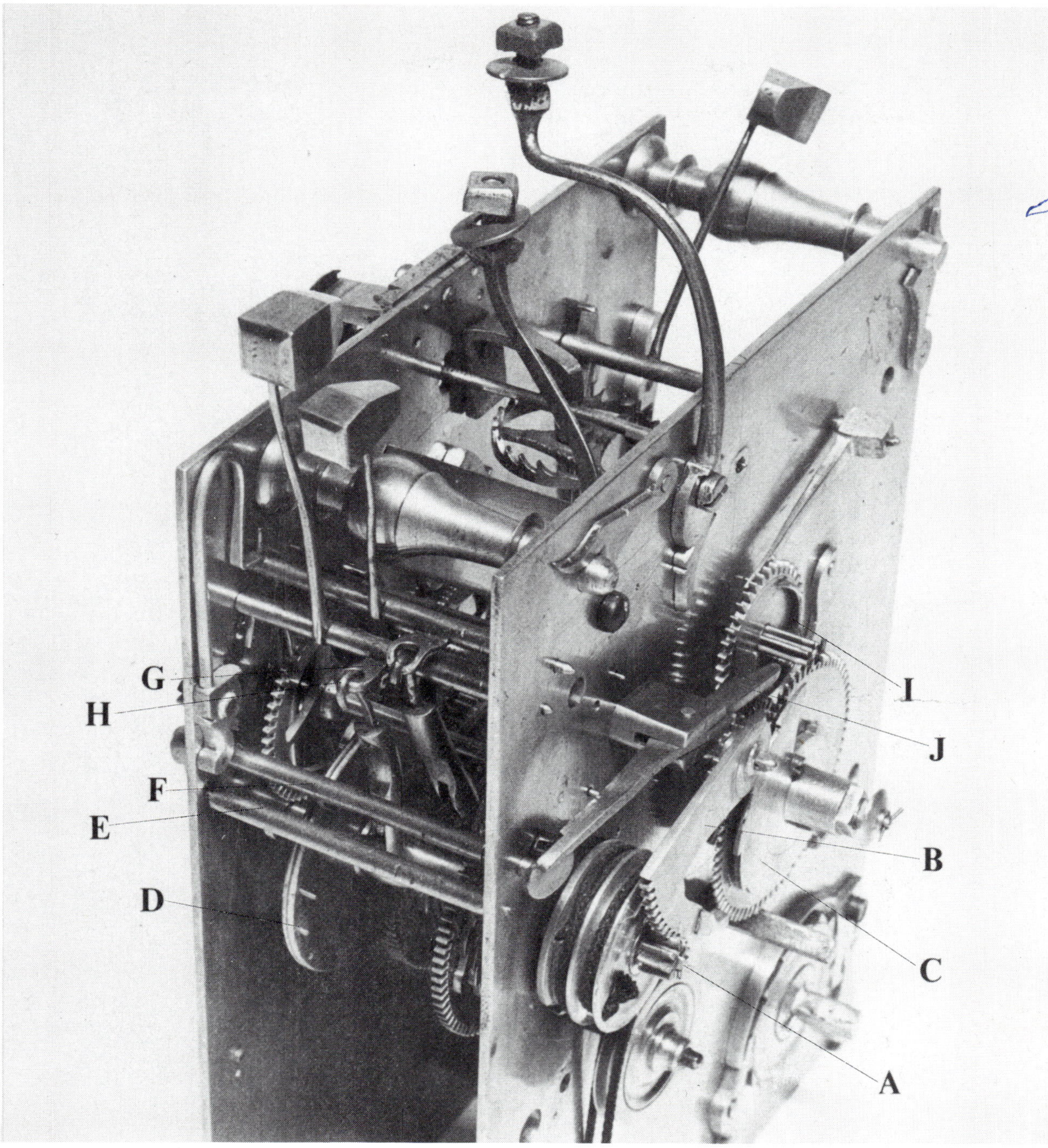

Joseph Knibb. On pulling either of the cords, the pinion A on the front of the upper pulley gathers the hour rack B on to its snail C (the lower pulley is simply a guide for the right hand cord) at the same time winding a spring housed in a fixed barrel under the upper pulley. The pulling of the cord turns the pin wheel D clockwise allowing hammer tails on the arbors E and F to ride over the pins without striking the bells. The open-ended connecting links G and H make this possible. The pin wheel has 12 hour pins on its back face and 3 quarter pins on its front. The number of hours which will be struck are determined by how far the hour rack tail is able to travel before being checked by a step on the hour snail, thereby allowing a greater or lesser number of hour pins to pass the hour hammer tail.

When the pull is released (there is no all-or-nothing piece) and as the spring unwinds, the hours are first sounded by those pins on the back of the pin wheel. As the spring further unwinds, the quarter pins sound the quarters by single blows on the quarter bell. The number of quarter blows struck are determined by the lateral position of the pumped hammer arbor as dictated by the quarter step snail I on the minute wheel via the rocking-bar (pump piece) J. It should be noted that the front pivot of the quarter hammer arbor F is kept continuously in contact with the rocking bar J (and thus with the quarter snail I) by a weak spring (not visible here; but see a similar one on page 68, top illustration and page 45 sketch) pressing from back to front. Thus, the quarter hammer arbor is pumped to and fro between the movement plates the whole time the clock is going. (In some systems the rocking bar only "senses" the quarter snail when the cord is pulled. See page 69 for an example). The above description is applicable to Knibb silent pulls as shown on pages 38-46.

Timepiece with Silent Pull sounding the Quarters on a Single Bell and with Alarm

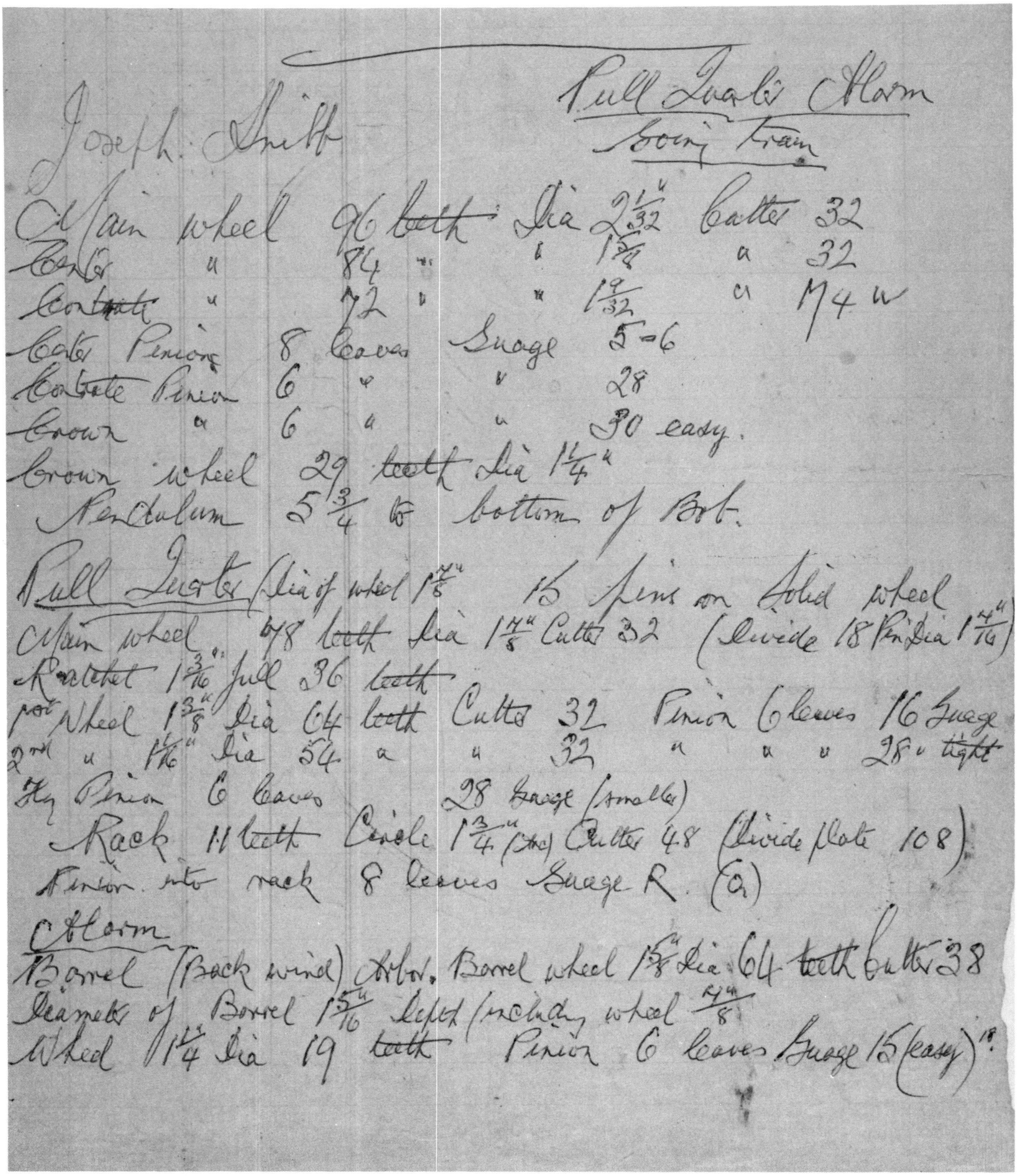
Joseph Knibb
Pull Quarter Alarm
Going train
Main wheel 96 ~~teeth~~ Dia 2 1/32" Cutter 32
Center " 84 " " 1 7/8 " 32
Contrate " 72 " " 1 9/32 " M4 w
Center Pinion 8 leaves Guage 5-6
Contrate Pinion 6 " " 28
Crown " 6 " " 30 easy.
Crown wheel 29 ~~teeth~~ dia 1 1/4"
Pendulum 5 3/4 to bottom of Bob.
Pull Quarter (dia of wheel 1 7/8" 15 pins on Solid wheel
Main wheel 48 ~~teeth~~ dia 1 7/8" Cutter 32 (Divide 18 Pin Dia 1 4/16")
Ratchet 1 3/16" full 36 ~~teeth~~
1st Wheel 1 3/8" dia 64 ~~teeth~~ Cutter 32 Pinion 6 leaves 16 Guage
2nd " 1 1/16" dia 54 " " 32 " " " 28 " tight
Fly Pinion 6 leaves 28 Guage (smaller)
Rack 11 ~~teeth~~ Circle 1 3/4" (Arc) Cutter 48 (Divide plate 108)
Pinion into rack 8 leaves Guage R. (G)
Alarm
Barrel (Back wind) Arbor, Barrel wheel 1 5/8" dia 64 ~~teeth~~ Cutter 38
Diameter of Barrel 1 5/16" depth (including wheel 5/8")
Wheel 1 1/4 dia 19 ~~teeth~~ Pinion 6 leaves Guage 15 (easy) 19

Joseph Knibb. Details of wheels and pinions in going, silent pull (referred to above as "Pull Quarter") and alarm trains. For a similar clock see photographs on pages 36 and 37.

Timepiece with Silent Pull sounding the Quarters on a Single Bell and with Alarm

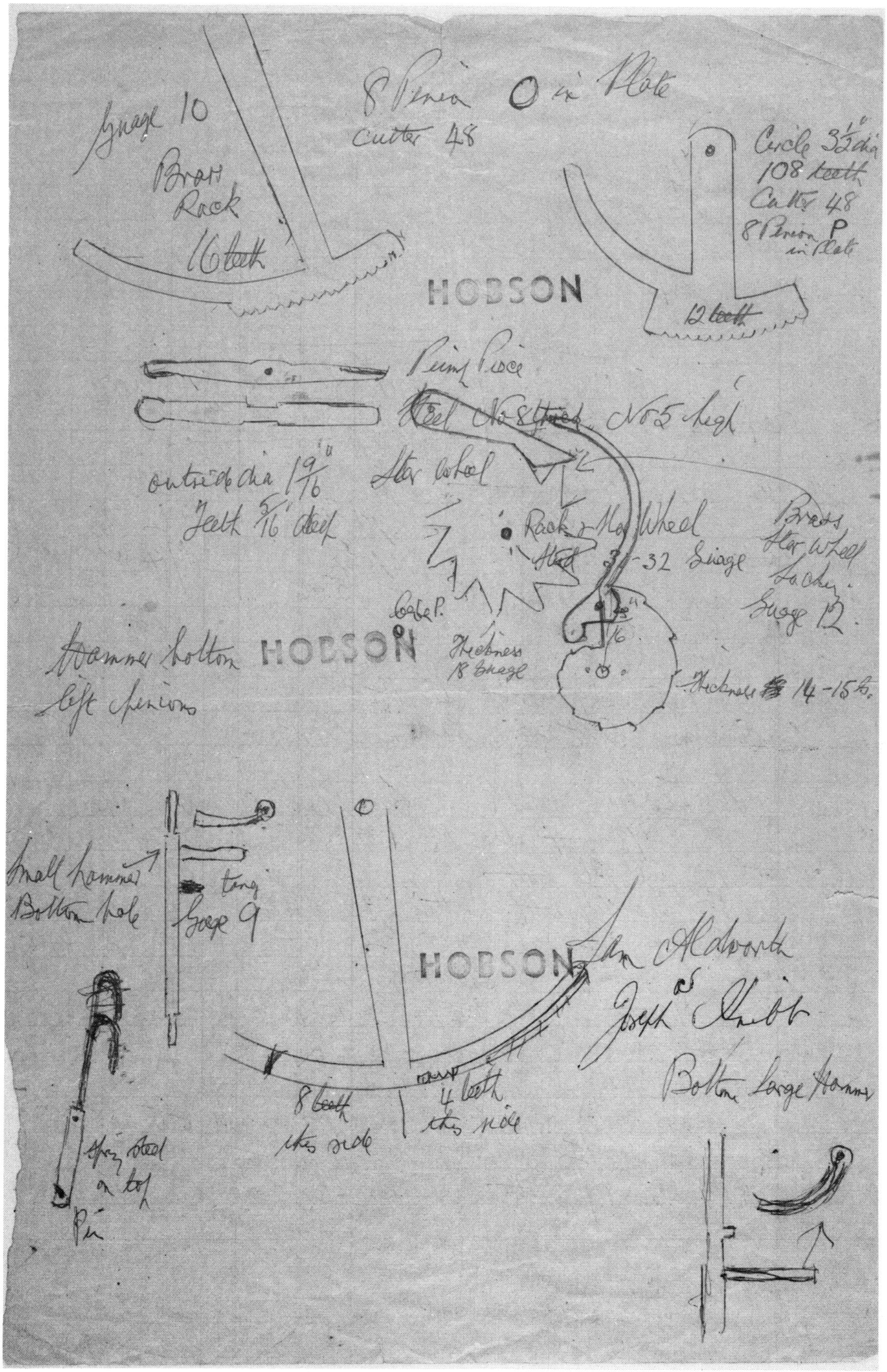

Joseph Knibb. Details of silent pull parts of clock on page 38. See also pages 36 and 37.

Timepiece with Silent Pull sounding the Quarters on a Single Bell

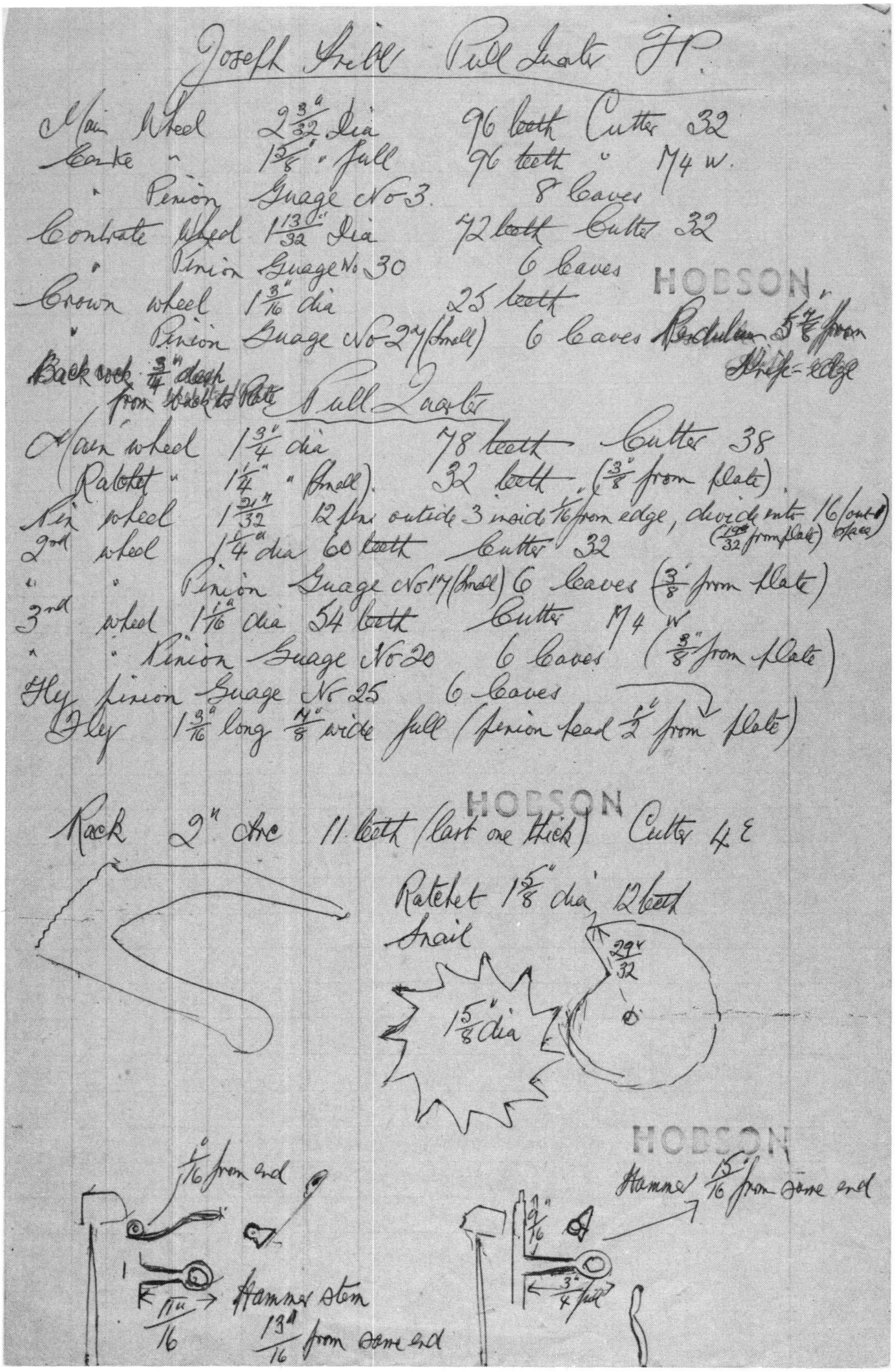
Joseph Knibb Pull Quarter T.P.

Main Wheel 2 3/32 dia 96 teeth Cutter 32
Centre " 1 5/8 " full 96 teeth " M4 w.
" Pinion Guage No 3. 8 Leaves
Contrate Wheel 1 13/32 dia 72 teeth Cutter 32
" Pinion Guage No 30 6 Leaves
Crown wheel 1 3/16 dia 25 teeth
" Pinion Guage No 27 (small) 6 Leaves Pendulum 5 7/8 from Knife-edge
Back cock 3/4" deep from back to plate

Pull Quarter

Main wheel 1 3/4 dia 78 teeth Cutter 38
Ratchet " 1 1/4" " (small). 32 teeth (3/8 from plate)
Pin wheel 1 21/32 12 pins outside 3 inside 1/16 from edge, divide into 16 (outside space) (19/32 from plate)
2nd wheel 1 1/4 dia 60 teeth Cutter 32
" " Pinion Guage No 17 (small) 6 leaves (3/8 from plate)
3rd wheel 1 1/16 dia 54 teeth Cutter M4 W
" " Pinion Guage No 20 6 Leaves (3/8 from plate)
Fly pinion Guage No 25 6 Leaves
Fly 1 3/16 long 7/8 wide full (pinion head 1/2 from plate)

HODSON

Rack 2" Arc 11 teeth (last one thick) Cutter 4 E

Ratchet 1 5/8 dia 12 teeth
Snail
29/32
1 5/8 dia

1/16 from end
Hammer 15/16 from same end
9/16
11/16
Hammer stem
13/16 from same end
3/4 full

Joseph Knibb. Dimensioned details of wheels and pinions, rack, snail and hammer work. See also pages 36 and 37.

Timepiece with Silent Pull sounding the Quarters on a Single Bell

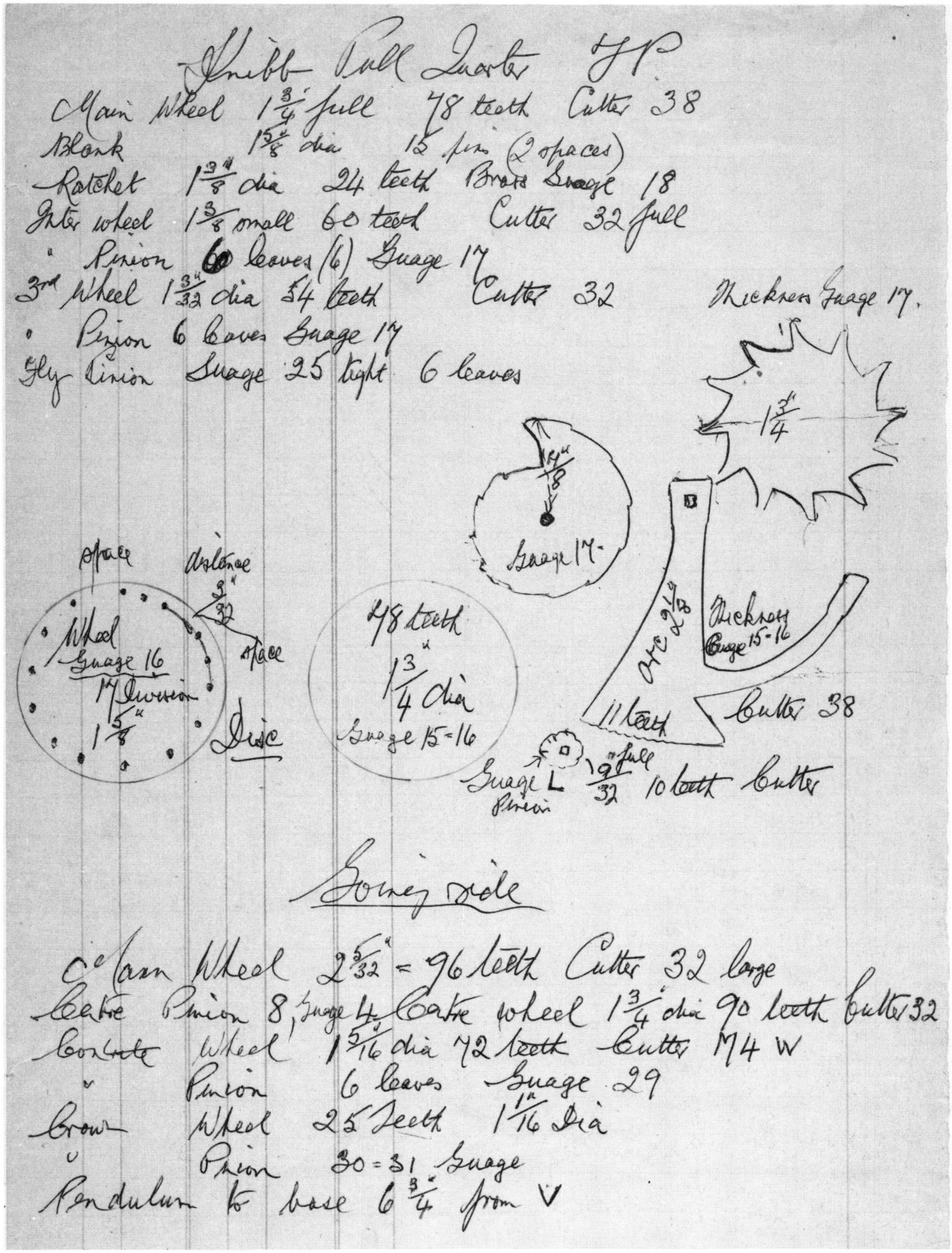
Knibb Pull Quarter T.P.

Main Wheel 1 3/4" full 78 teeth Cutter 38
Blank 1 5/8" dia 12 pins (2 spaces)
Ratchet 1 3/8" dia 24 teeth Brass Guage 18
Inter wheel 1 3/8 small 60 teeth Cutter 32 full
" Pinion 6 leaves (6) Guage 17
3rd Wheel 1 3/32" dia 54 teeth Cutter 32 Thickness Guage 17.
" Pinion 6 leaves Guage 17
Fly Pinion Guage 25 tight 6 leaves

Going side

Main Wheel 2 5/32" = 96 teeth Cutter 32 large
Centre Pinion 8, Guage 4, Centre wheel 1 3/4" dia 90 teeth Cutter 32
Contrate Wheel 1 5/16" dia 72 teeth Cutter M4 W
" Pinion 6 leaves Guage 29
Crown Wheel 25 teeth 1 1/16" Dia
" Pinion 30-31 Guage
Pendulum to base 6 3/4" from V

Joseph Knibb. Details of going and silent pull wheel and pinion work, together with dimensioned sketches showing silent pull main wheel, hour snail and its star wheel and hour rack. See also pages 36 and 37.

Timepiece with Silent Pull sounding the Quarters on a Single Bell and with Alarm

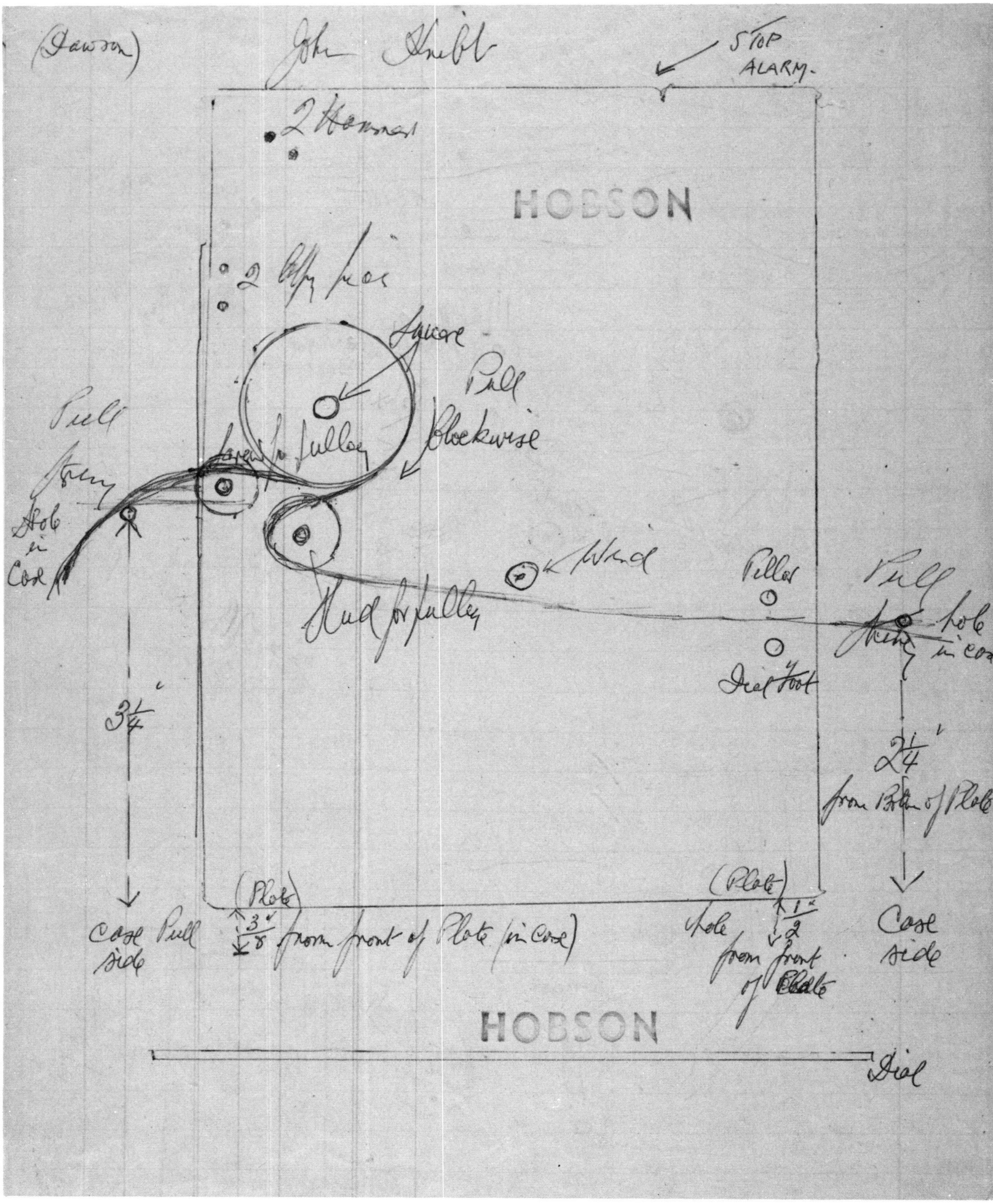

John Knibb. View of front plate showing pulley work acting in either direction, positions of hammer work and dimensions relating movement to clock case. See also pages 36 and 37.

Timepiece with Silent Pull sounding the Quarters on a Single Bell and with Alarm

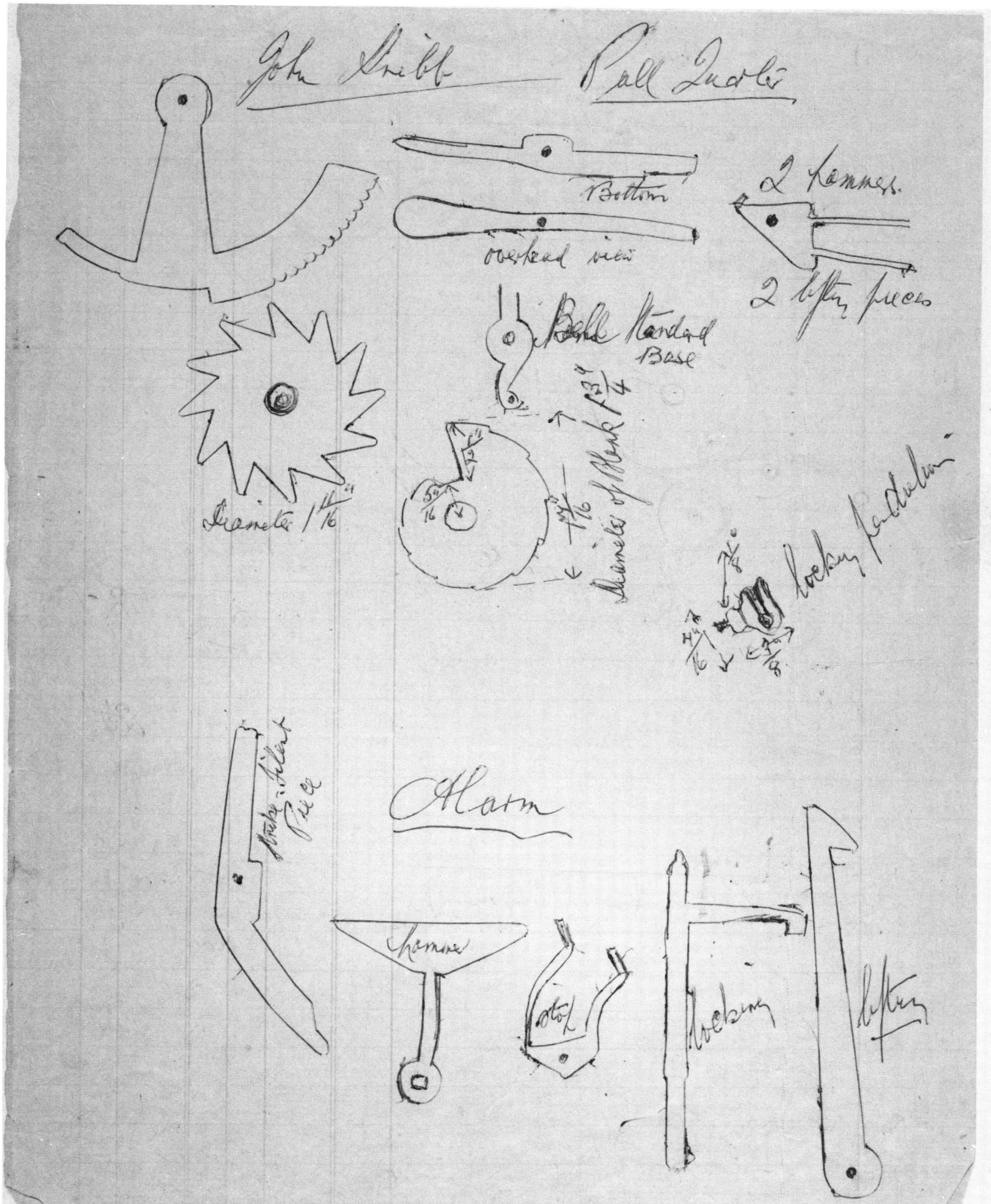

John Knibb. Detail sketches made from silent pull and alarm work of clock outlined on page 42. See also pages 36 and 37.

Timepiece with Silent Pull sounding the Quarters on a Single Bell

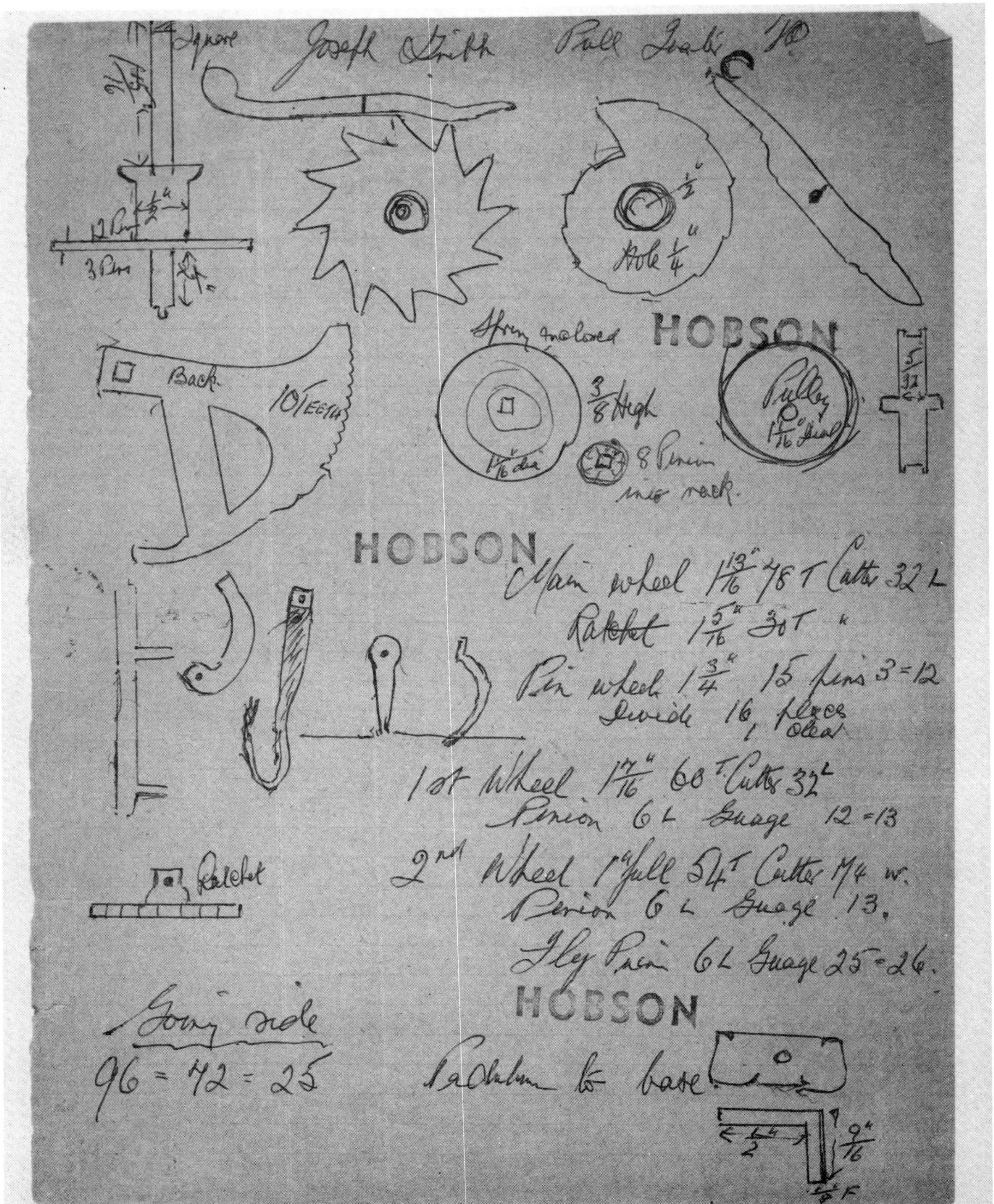

Joseph Knibb. Details and dimensions of silent pull wheelwork and other parts. To be used in conjunction with pages 36 and 37.

Timepiece with Silent Pull sounding the Quarters on a Single Bell

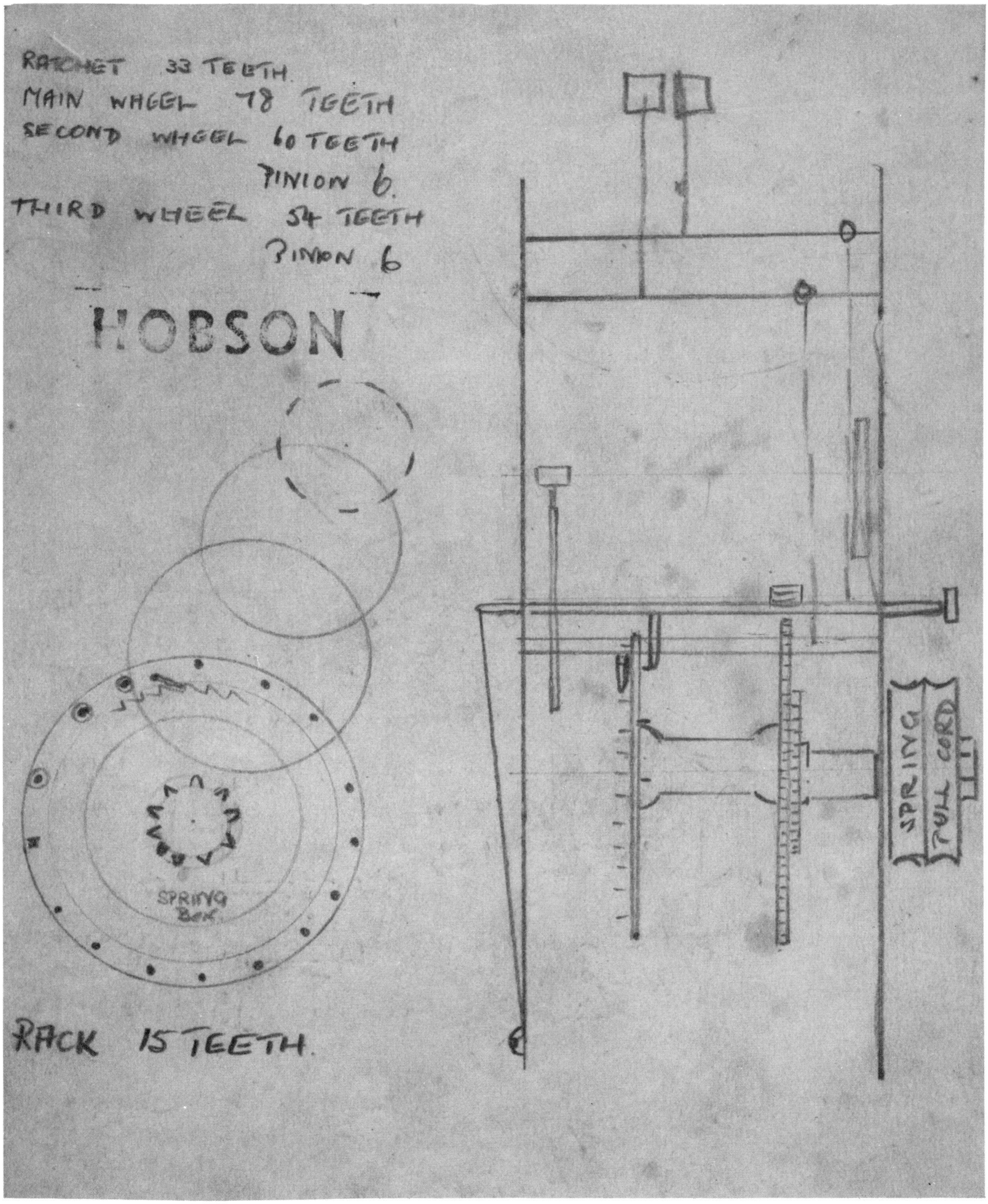

John Knibb. Side view of silent pull wheel and hammer work, together with train. See pages 36 and 37.

Timepiece with Silent Pull sounding the Quarters on a Single Bell

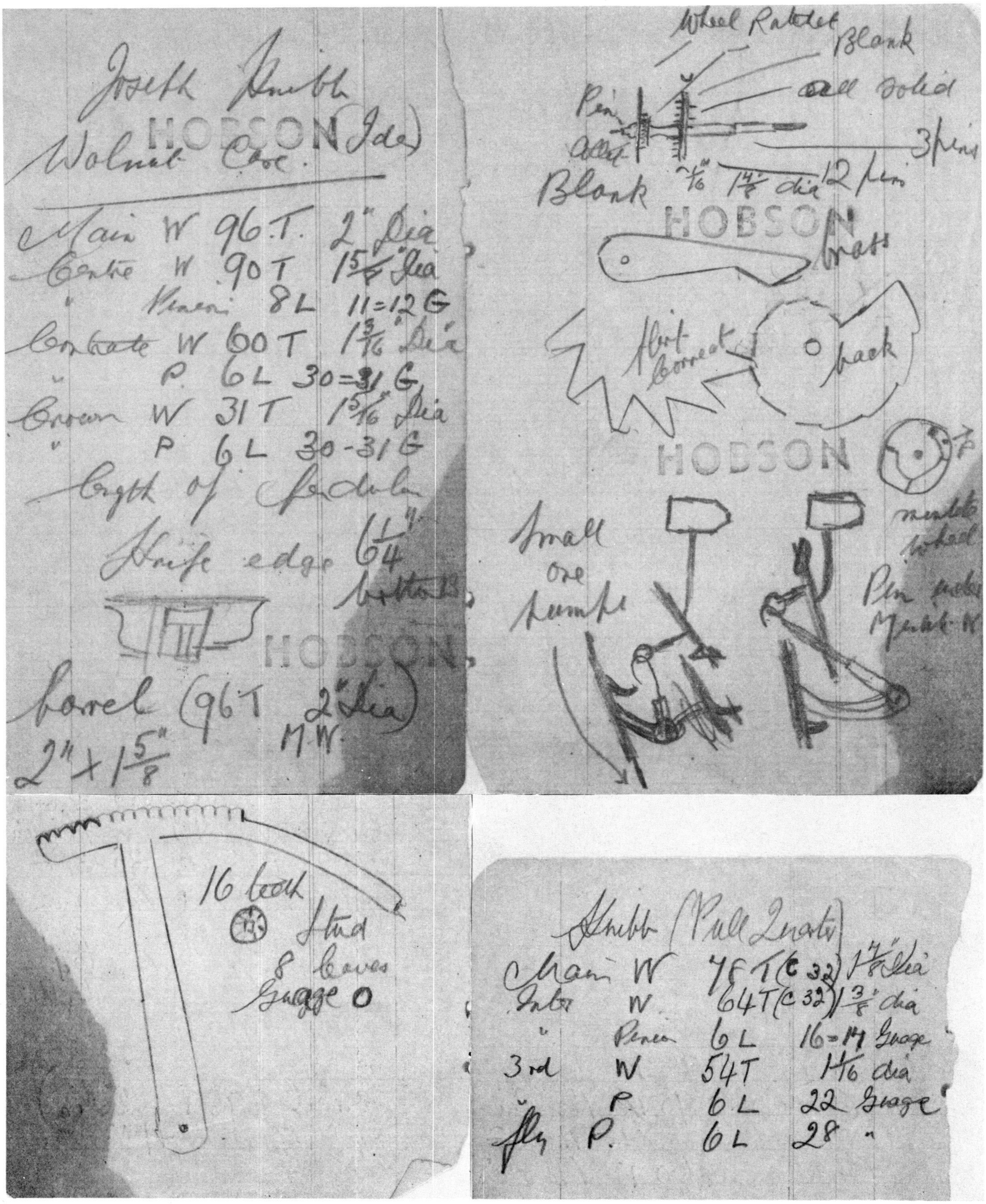

Joseph Knibb. The "Iden" clock in walnut case. Details of going train and silent pull. See also pages 36 and 37.

Striking Clock with Pull Quarter on a Single Bell

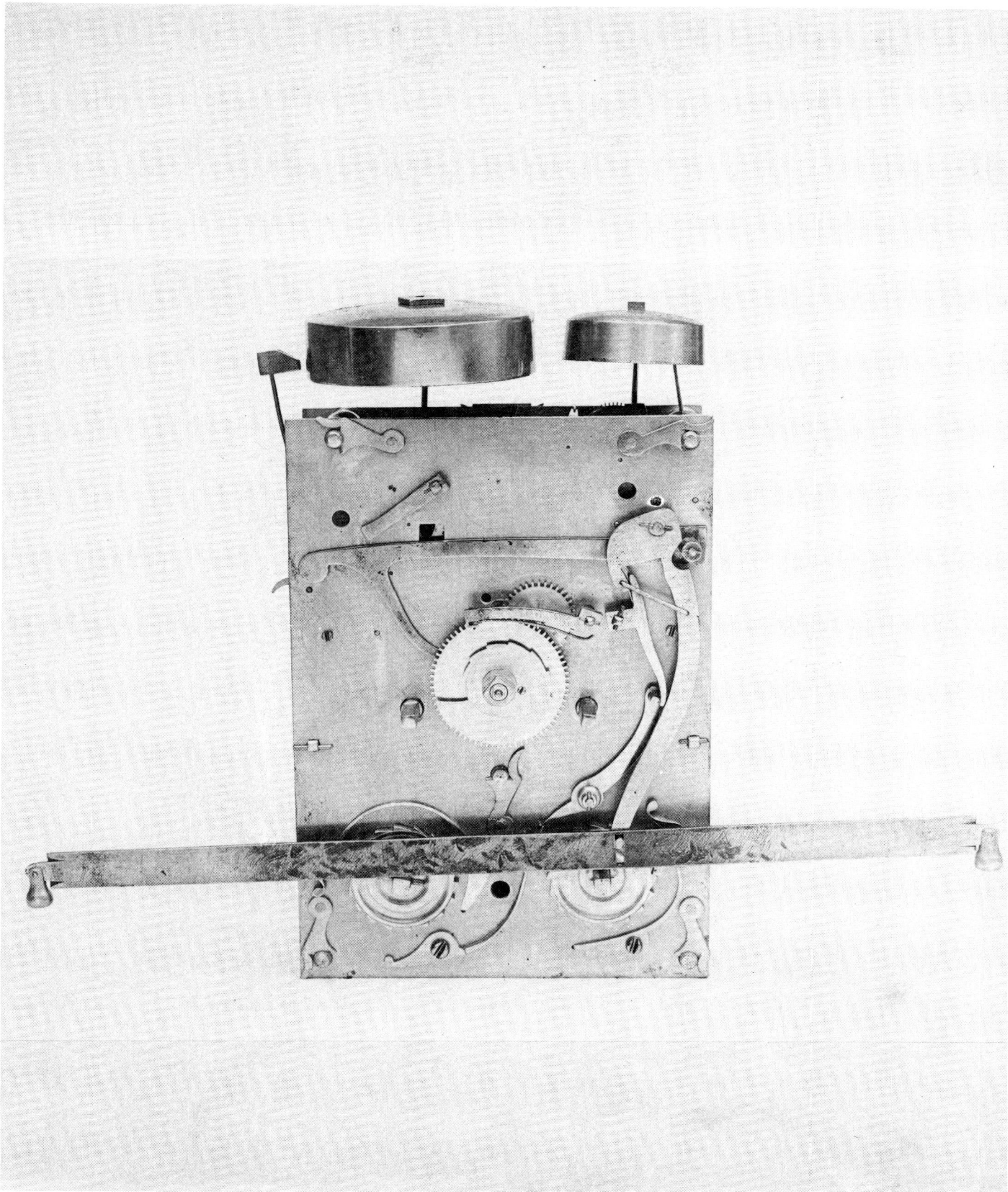

Joseph Knibb. View of front plate showing pull quarter and striking work and now-rare steel pull bar. Lee *Knibb*, plate 80, shows such a clock. These bars were often replaced by side cords and guide pulley boxes as latterly used by Knibb himself. See Lee, plate 84. Knibb timepieces with silent pull do not use the pull bar arrangement because it does not afford sufficient movement. Knibb's bar was sometimes used by Henry Jones. See page 51 for another form of Knibb pull suitable for pull quarter.

Striking Clock with Pull Quarter on a Single Bell

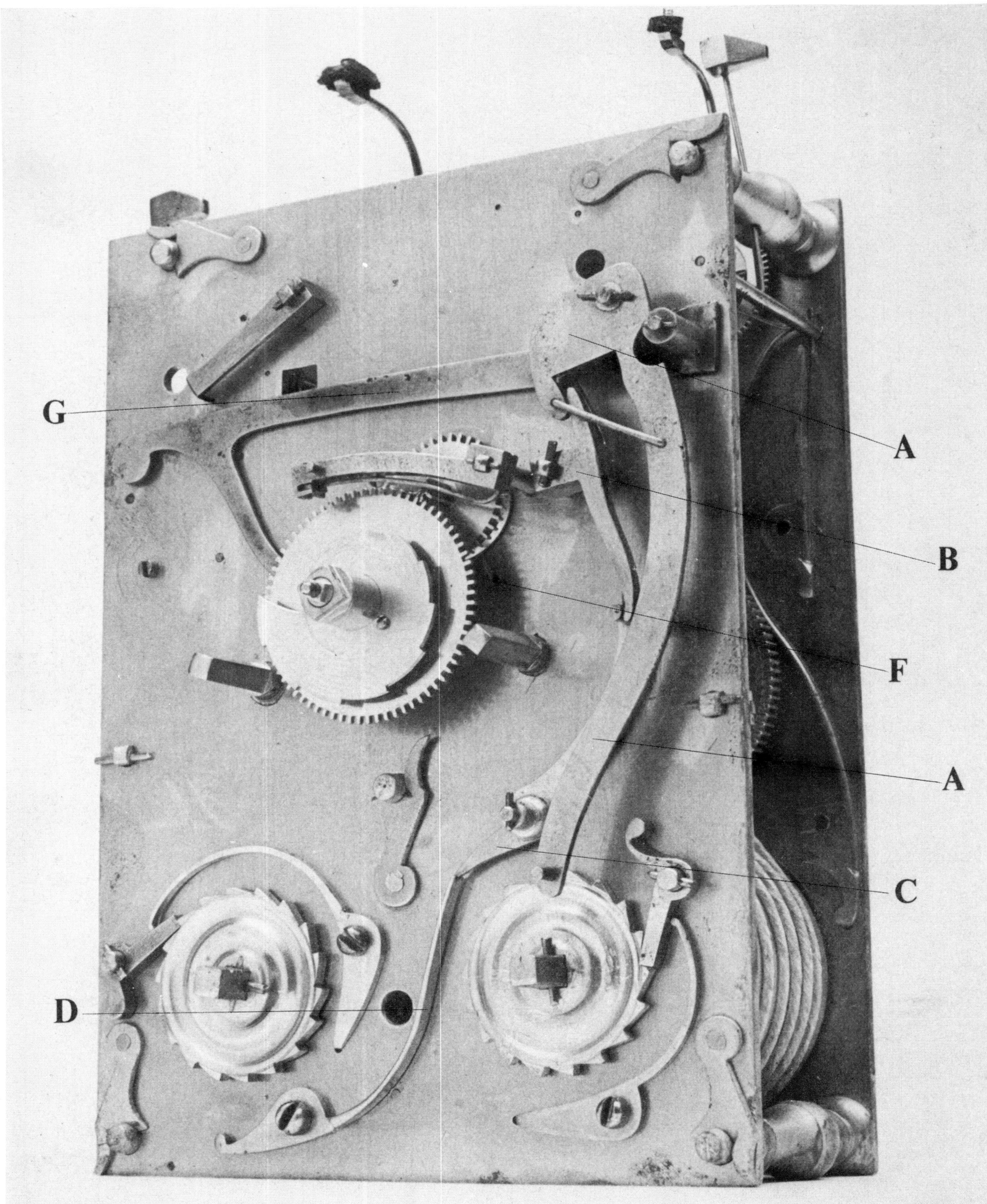

Joseph Knibb. Another view of the clock on page 47. Note the absence of an hour wheel bridge. Knibb sometimes used one and sometimes not. In the latter case the hour wheel rides on the cannon.

Striking Clock with Pull Quarter on a Single Bell

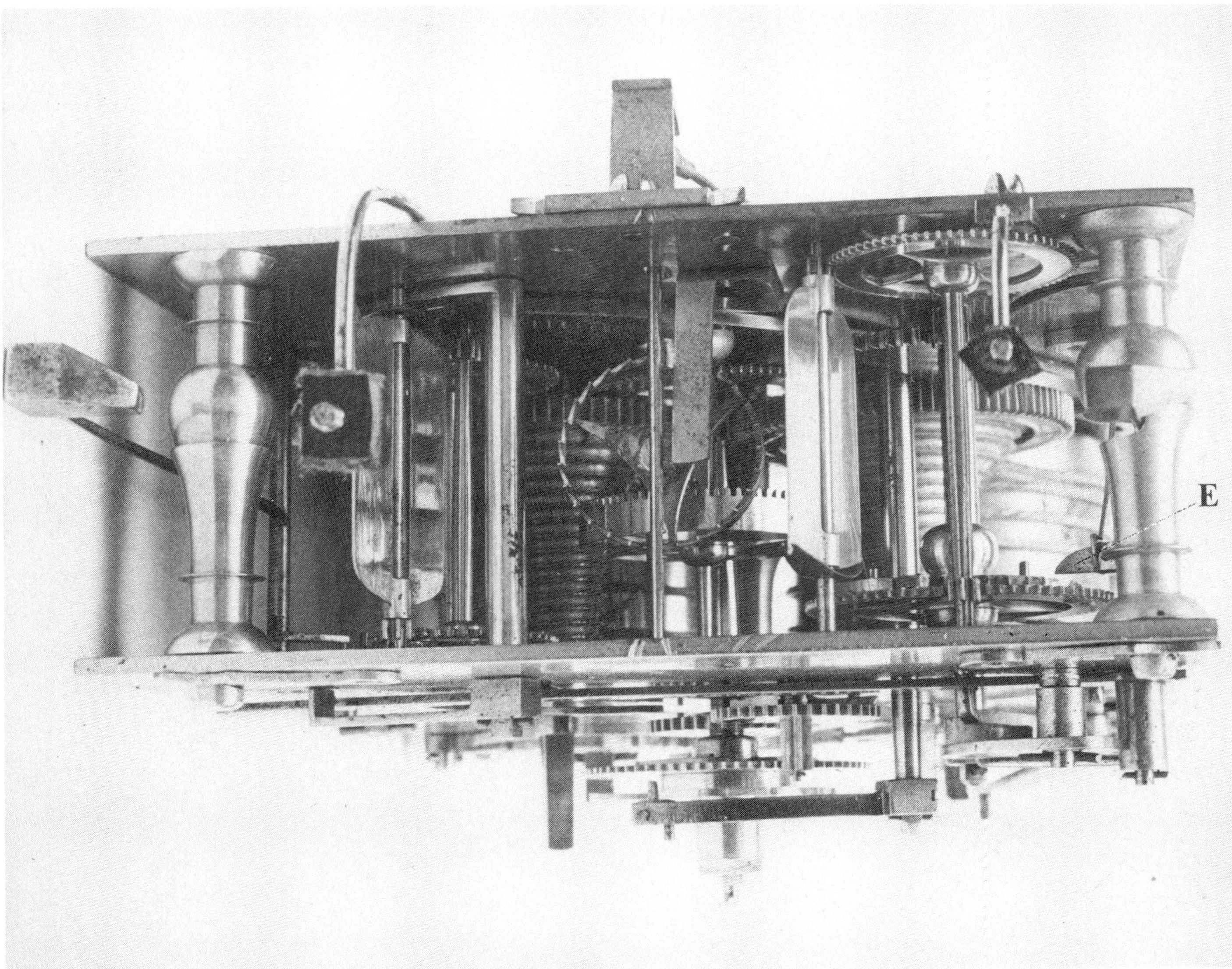

Joseph Knibb. Please see also pages 47 and 48. On pulling the bar in either direction, the vertical lever AA rotates lever B in a clockwise direction thereby pushing lever C against the brass blade spring D which accordingly is cocked. The lever B is squared on to the internal quarter rack arbor which, on pulling off, gathers the main pinion and clickwheel, the quarter pins passing freely over the fly-back hammer tail E. The number of quarters which will be struck are determined by the quarter rack tail F which passes through a slot in the front plate in order to reach the quarter snail behind the minute wheel. At the same time the lever B, which is bi-furcated at its top, lifts the horizontal lever G which in turn releases and holds up the hour striking train. As the pull is released, spring D turns the train, sounding the quarters and ultimately lowering lever G in order to let off the hours. The curious wire strap at the top of lever AA is purely a strengthening piece as originally used by Knibb. The delicacy and refinement of Knibb's work is well demonstrated by the fact that he was able to drive his pulled-quarters by such a thin brass spring. Compare with clocks shown elsewhere.

Striking Clock with Pull Quarter on a Single Bell

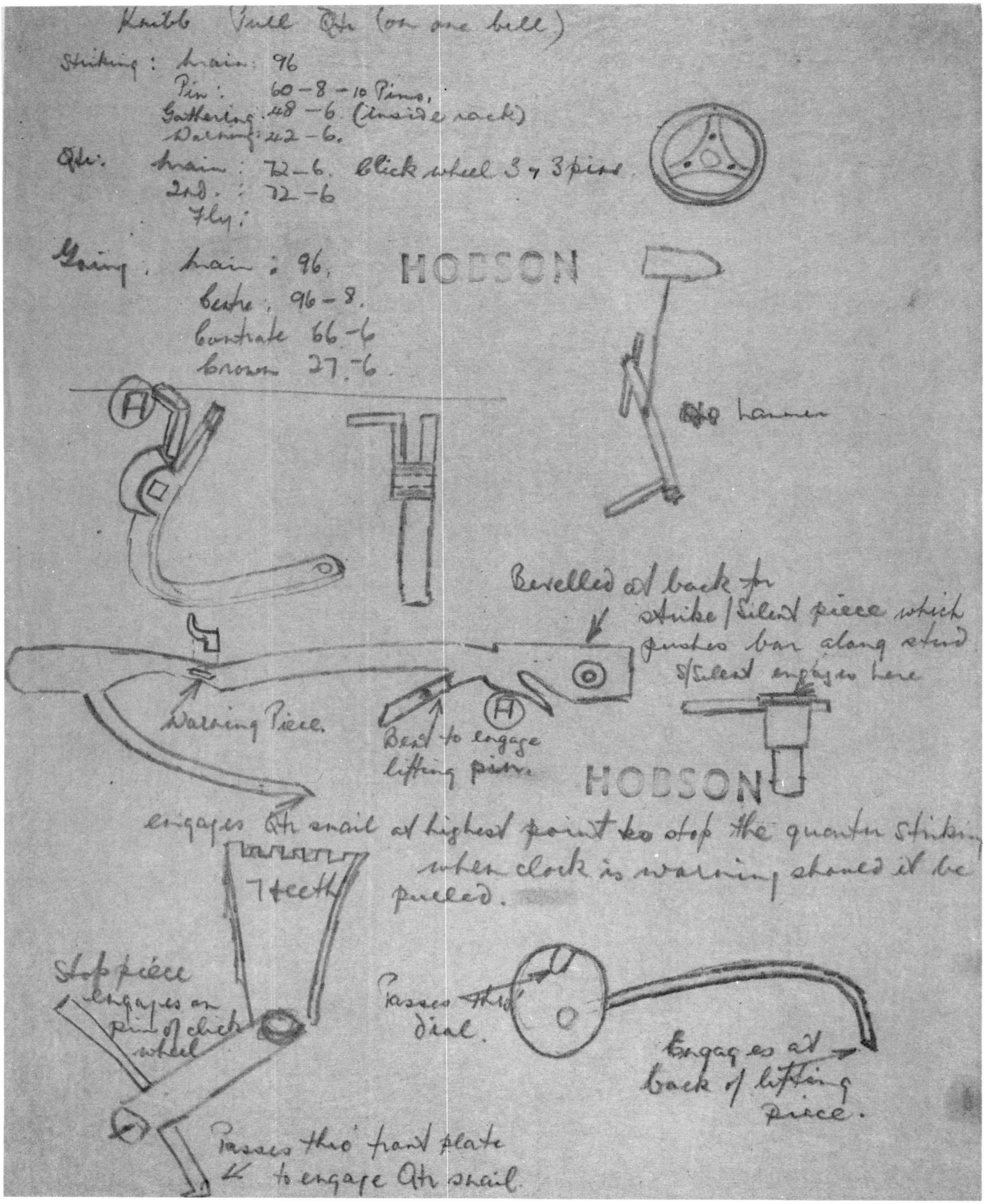

Joseph Knibb. These sketches should be studied in conjunction with the photographs on pages 47-49.

Striking Clock with Pull Quarter on a Single Bell

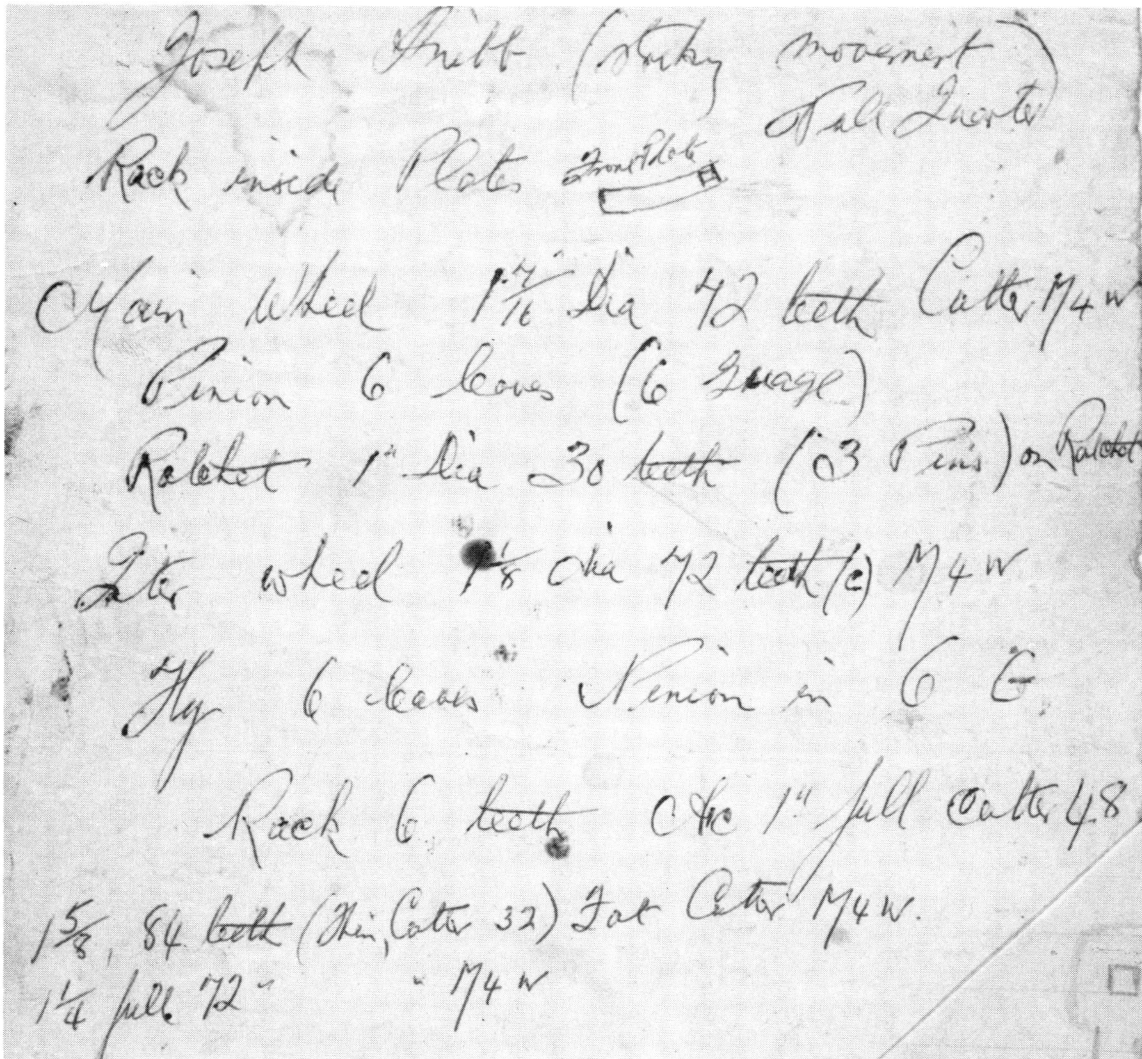
Joseph Knibb (Striking movement)
Pull Quarter
Rack inside Plate Front Plate
Main Wheel 1 7/16 Dia 72 teeth Cutter M4W
Pinion 6 Leaves (6 Guage)
Ratchet 1" Dia 30 teeth (3 Pins) on Ratchet
Qtr wheel 1 1/8 Dia 72 teeth (c) M4W
Fly 6 Leaves Pinion in 6 G.
Rack 6 teeth Dia 1" full Cutter 4.8
1 5/8, 84 teeth (Thin Cutter 32) Fat Cutter M4W.
1 1/4 full 72" " M4W

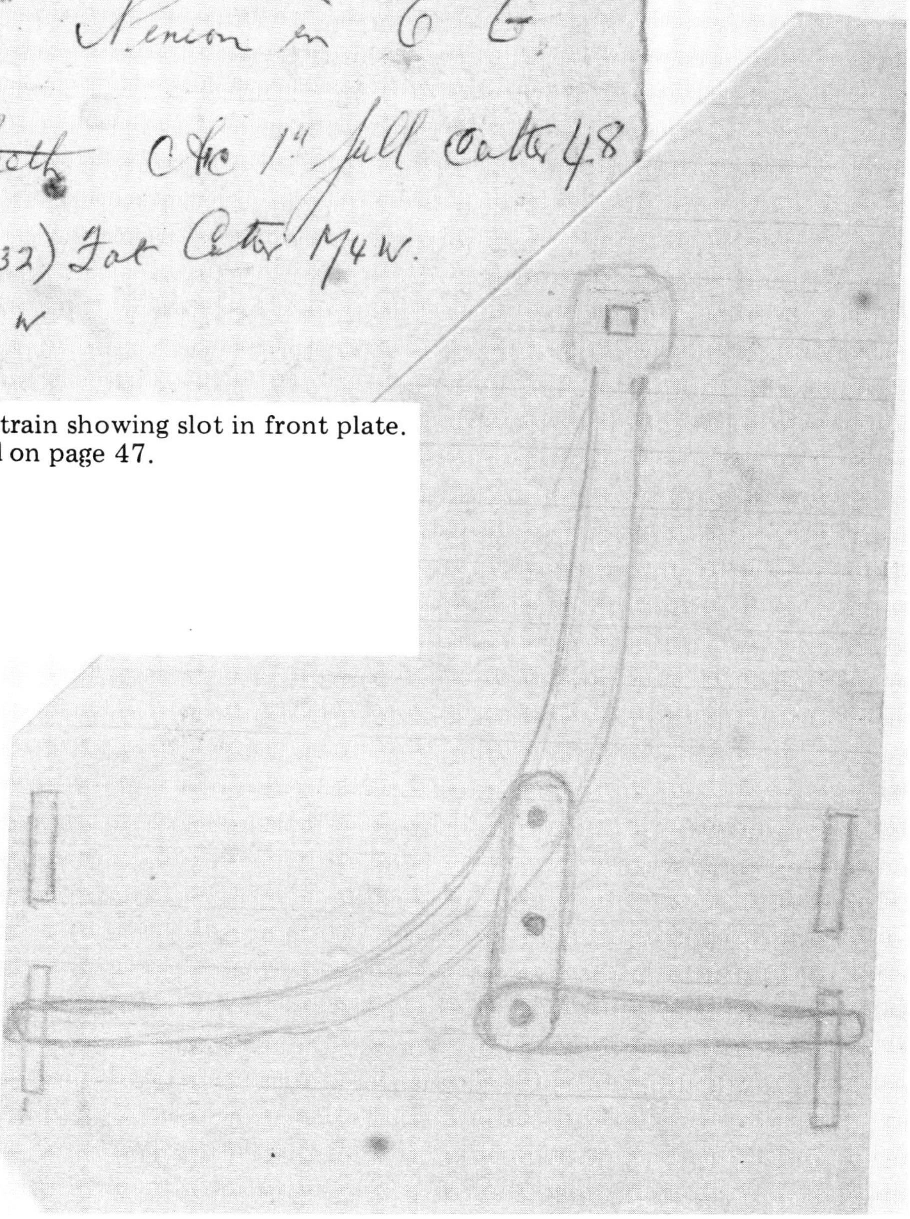

Joseph Knibb. Details of pull quarter train showing slot in front plate. The sketch bottom right is mentioned on page 47.

Striking Clock with Pull Quarter on a Single Bell

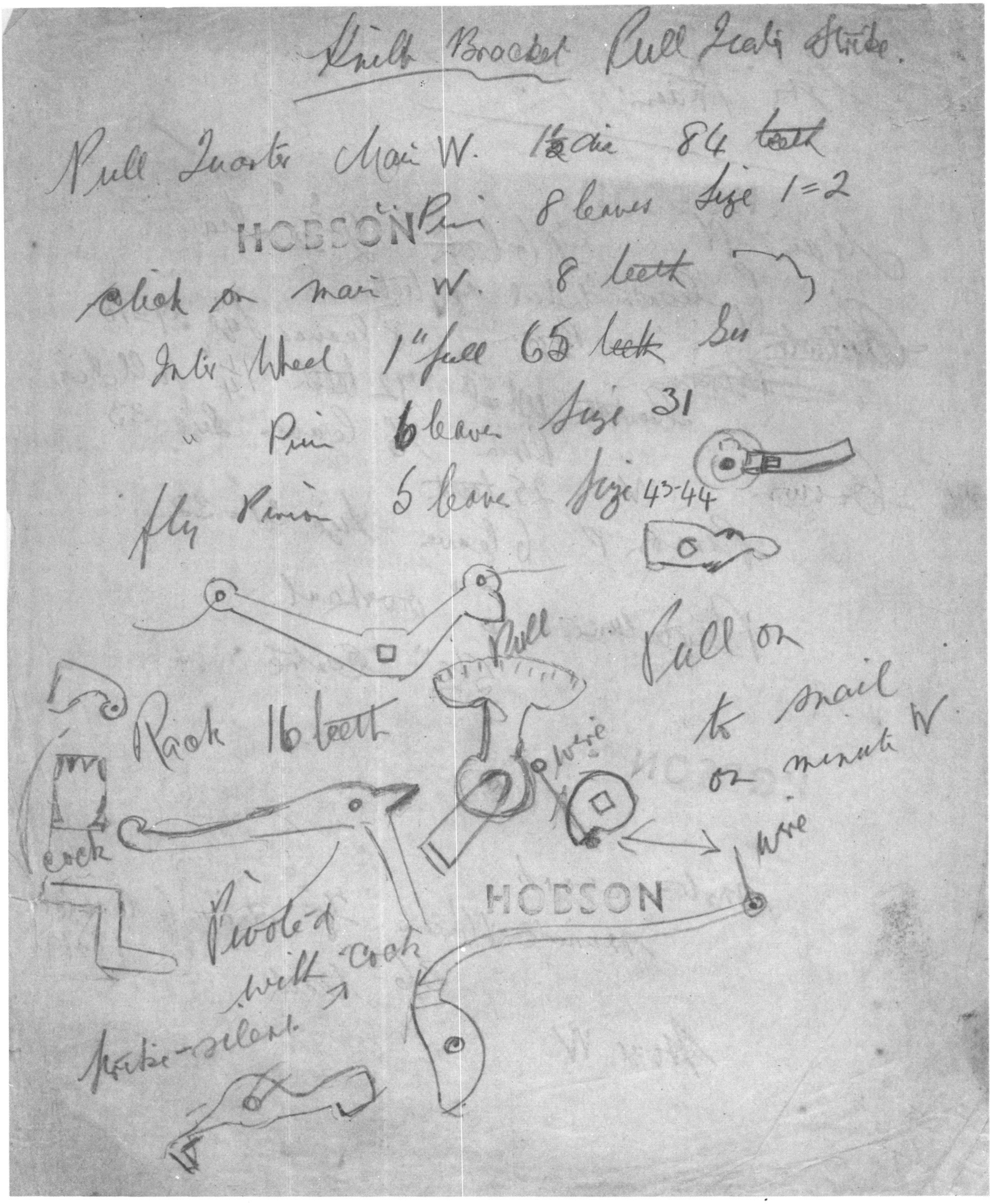

Joseph Knibb. Details of repeating parts, strike/silent lever and pull quarter train. In the centre of the page note the unusual pull lever worked from either side. This lever is positioned vertically on the back plate of the clock. See Lee *Knibb*, plate 132, in which the cords should be shown going in opposite directions.

Striking Clock with Pull Quarter on Two Bells

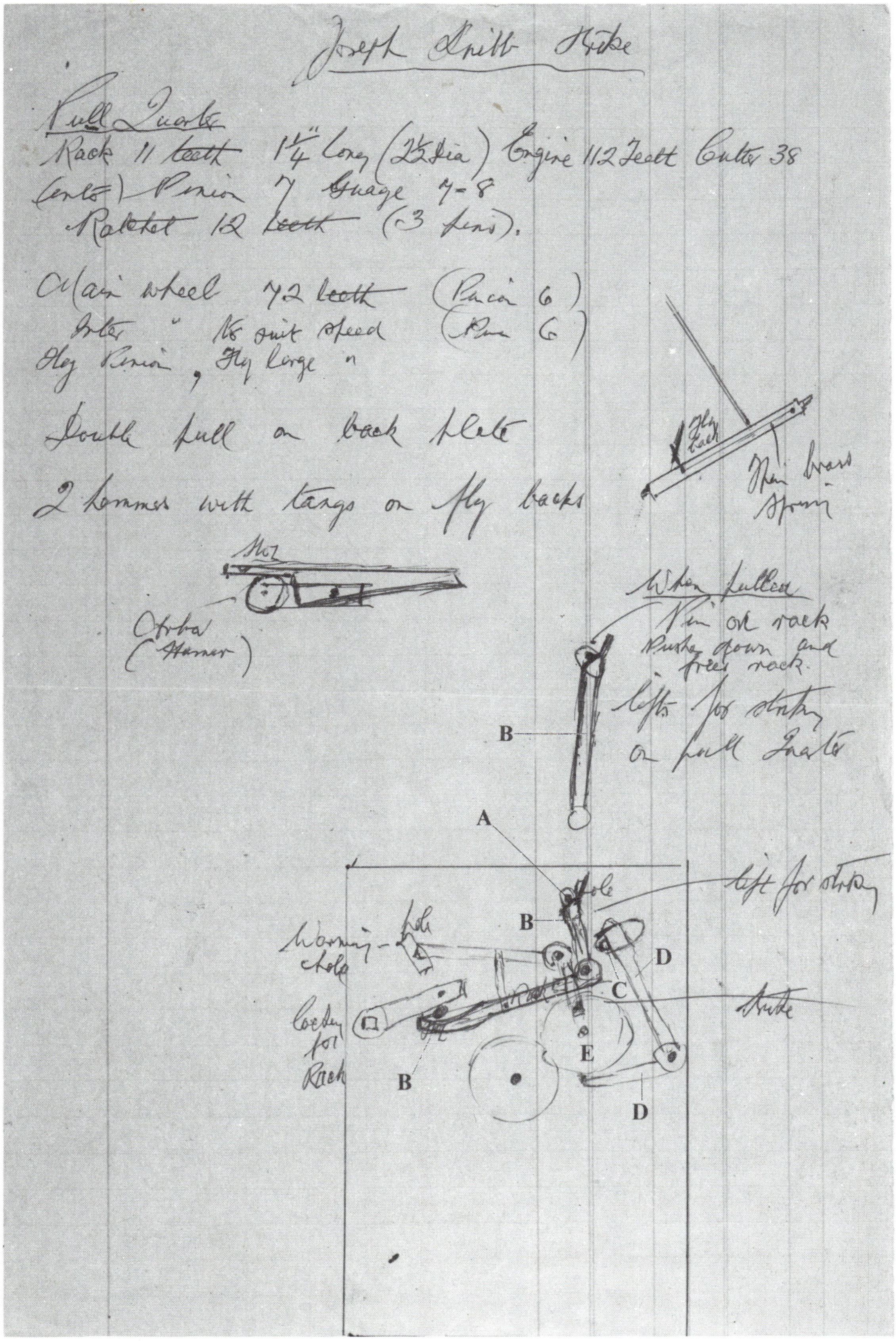

Joseph Knibb. Incomplete notes and sketches. This shows an alternative but rarely used system. The internal pull quarter work is the same as that shown on pages 47-49 other than that the quarter rack has two tails. The upper tail A passes through a slot in the front plate. When the double lever on the back plate is pulled, A moves downwards causing L-shaped lever BB to release and warn the hour striking train. At the same time the lower tail C pushes against lever DD causing its tail to move upwards as far as is allowed by the quarter snail which lies behind the minute wheel E.

Striking Clock with Pull Quarter on Three Bells

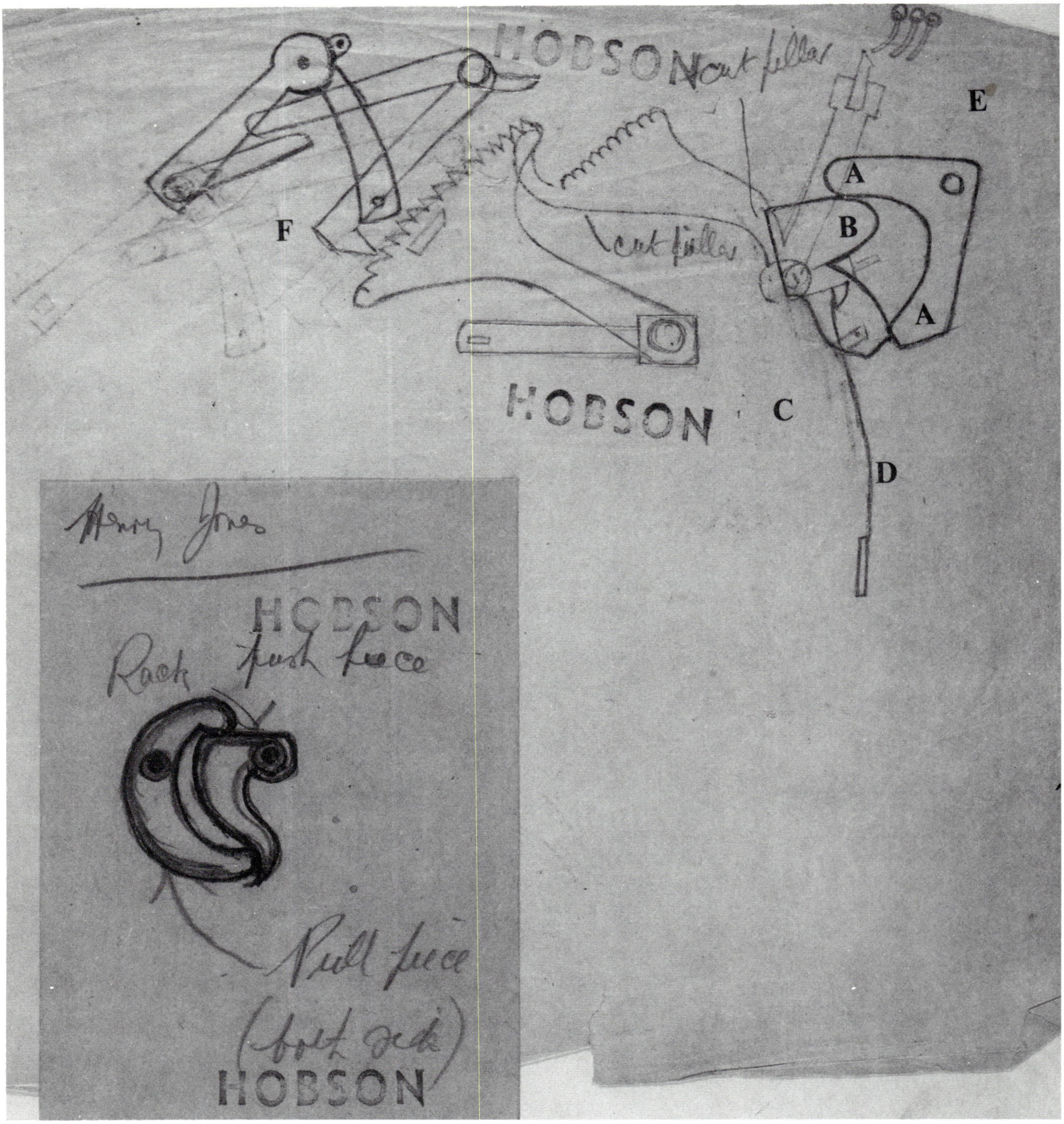

On pulling off, the lever AA, squared on the arbor carrying the pull lever, rotates in either direction, thereby always causing B to rotate clockwise. The movement of B is controlled by the quarter snail (not shown) upon which the block C impinges. The blade spring D drives the pull quarter. The fly back hammer lifting piece E leaves the three hammer tails undisturbed during the pull. It is not clear from these sketches how the hour work is set off after the quarters have been sounded. Probably the upward-curved tail of the quarter rack raises an internal horizontal lever (not shown) which in turn is attached to the same arbor as lever F.

Henry Jones. Unlike Tompion and Knibb, Jones used a variety of pull-repeating mechanisms. His ideas developed during his working life without reaching either the perfection of Tompion or the simplicity of Knibb. This very interesting sketch is drawn as though the front plate were made of glass showing simultaneously the motions both in front and behind the plate. The sketch also shows the internal striking rack used by Jones.

Striking Clock with Pull Quarter on Three Bells

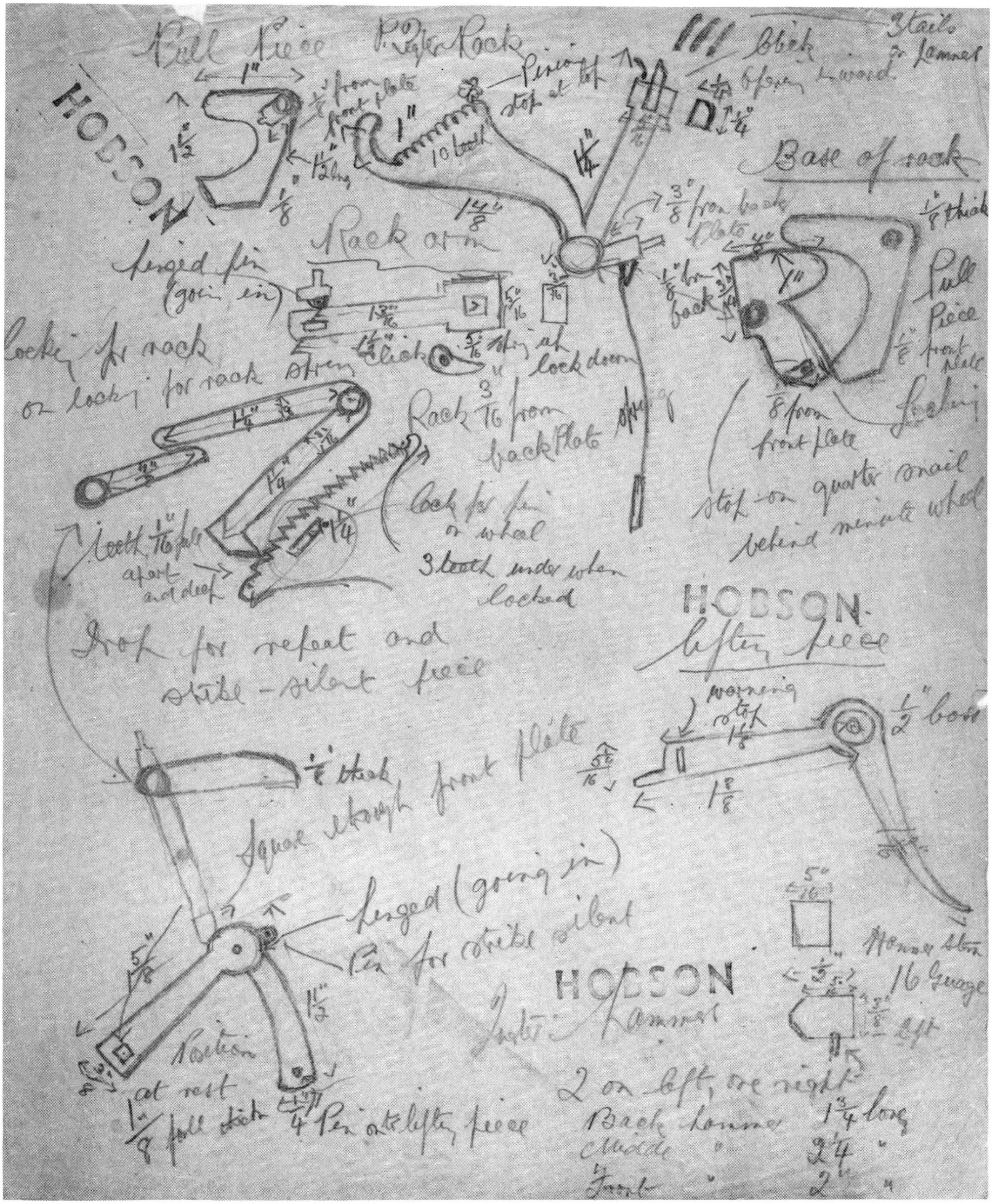

Henry Jones. Details of individual parts of the pull quarter work shown opposite with dimensions and brief descriptions.

Striking Clock with Pull Quarter on Three Bells

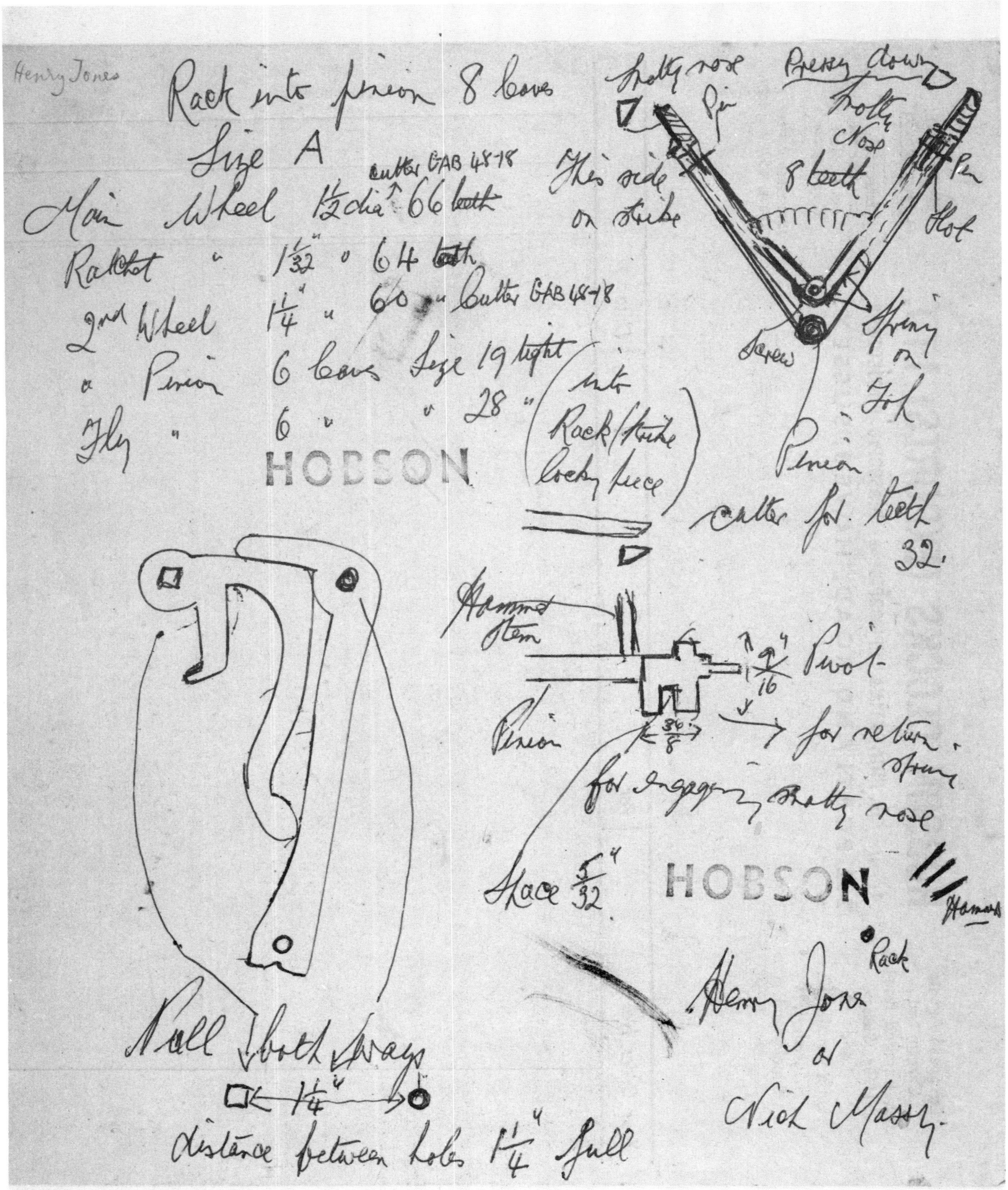

Henry Jones. Details of internal quarter rack, fly-back hammer lifting pieces ("snotty nose"), two way pull piece and pull quarter train. This system was sometimes used by Nicholas Massy and by James Markwick.

Striking Clock with Pull Quarter on Six Bells and with Alarm

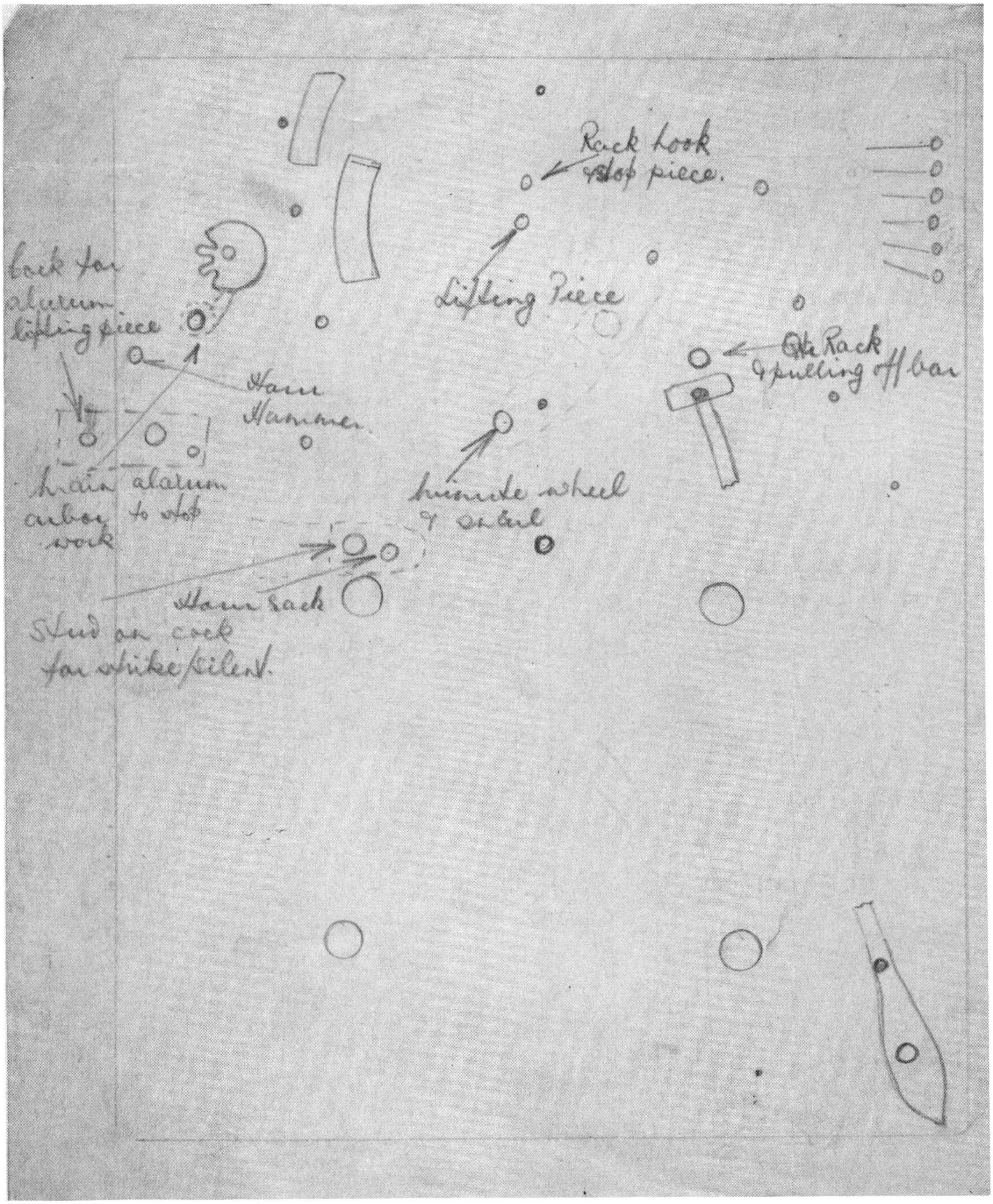

Jacob Massey. This early six-bell pull quarter works on the internal quarter rack and pinion principle. The pins on the pin wheel not only lift the hammers but also provide locking after the quarters have been sounded. Note that in this clock the hour rack has migrated from between the plates to the front plate. Probable date, circa 1710.

Striking Clock with Pull Quarter on Six Bells and with Alarm

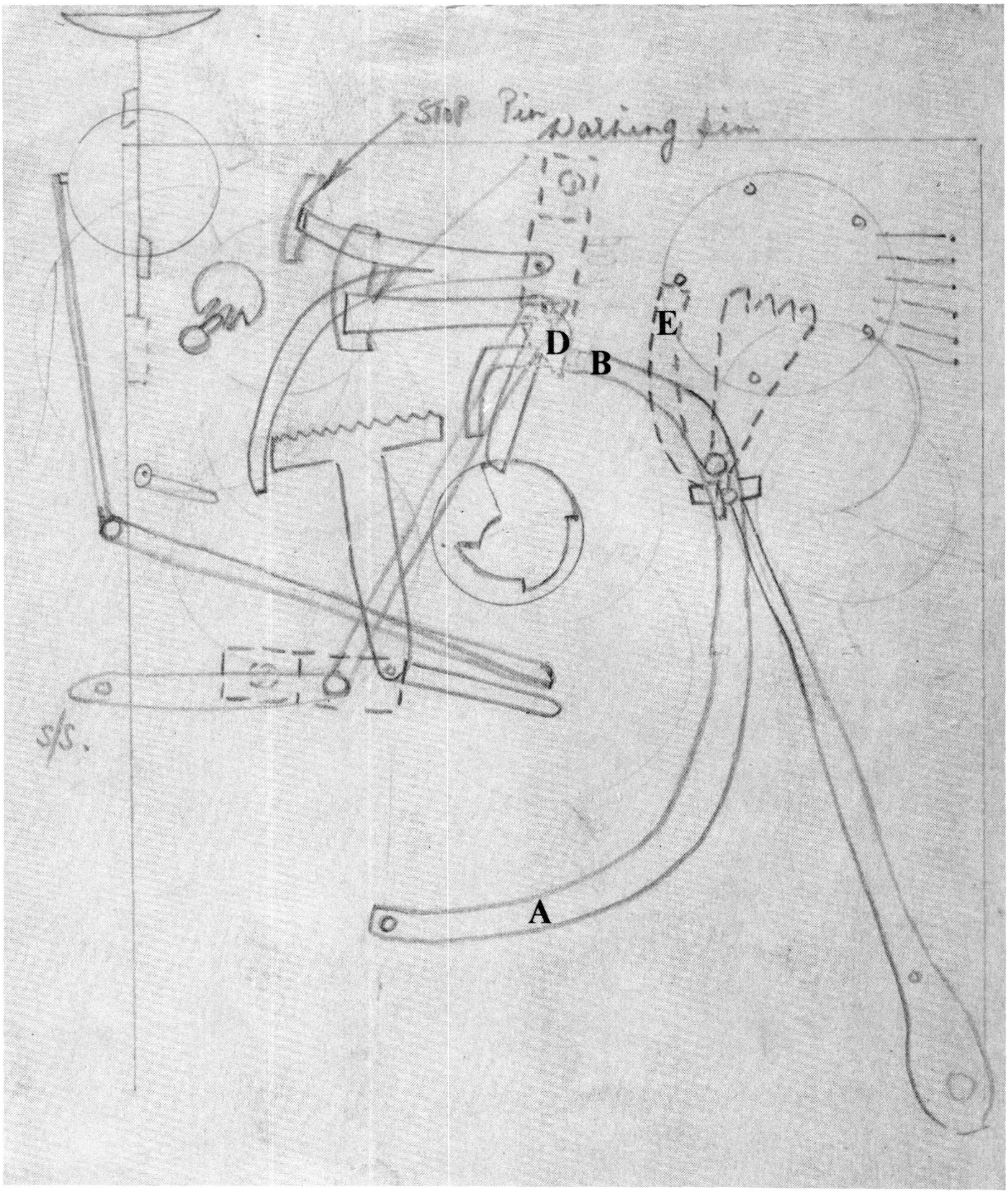

Jacob Massey. Lever A is squared on to the internal quarter rack (dotted). When A is pulled to the right the quarter tail B is checked by the quarter snail. The quarter tail rotates the main pinion and clickwheel (not shown) clockwise leaving the pin wheel stationary. At the same time a small block C on the back of B (see also detail sketches on page 59) presses down projection D thereby releasing and warning the hour striking train. As the rack returns it drives the pin wheel through its clickwork so sounding the quarters until lever E integral with the quarter rack blocks the path of one of the pins. The return of lever B to its rest position allows the hour striking to take place.

Striking Clock with Pull Quarter on Six Bells and with Alarm

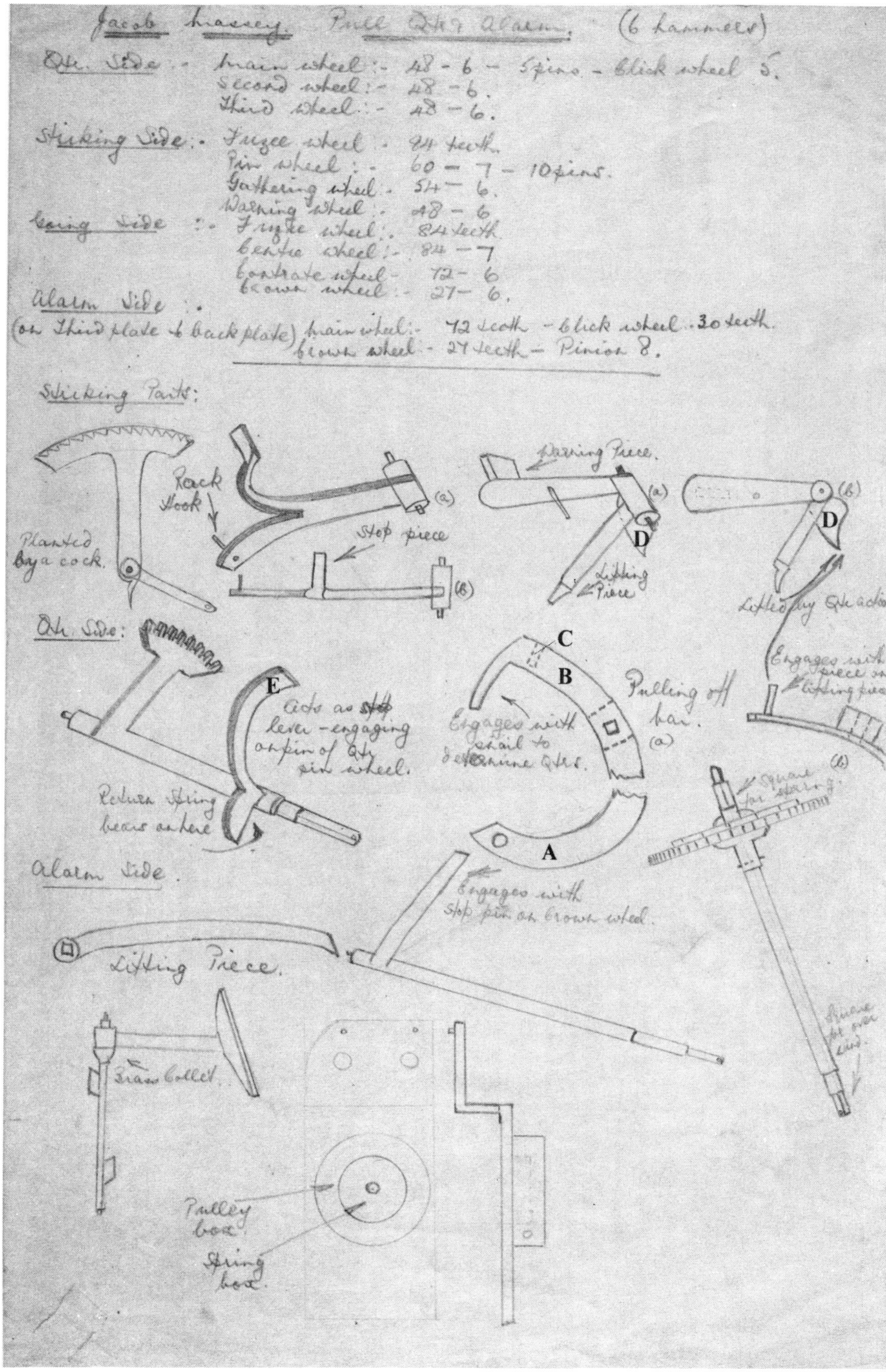

Jacob Massey. Train counts with sketches of parts shown on page 58.

Striking Clock with Pull Quarter on Three Bells

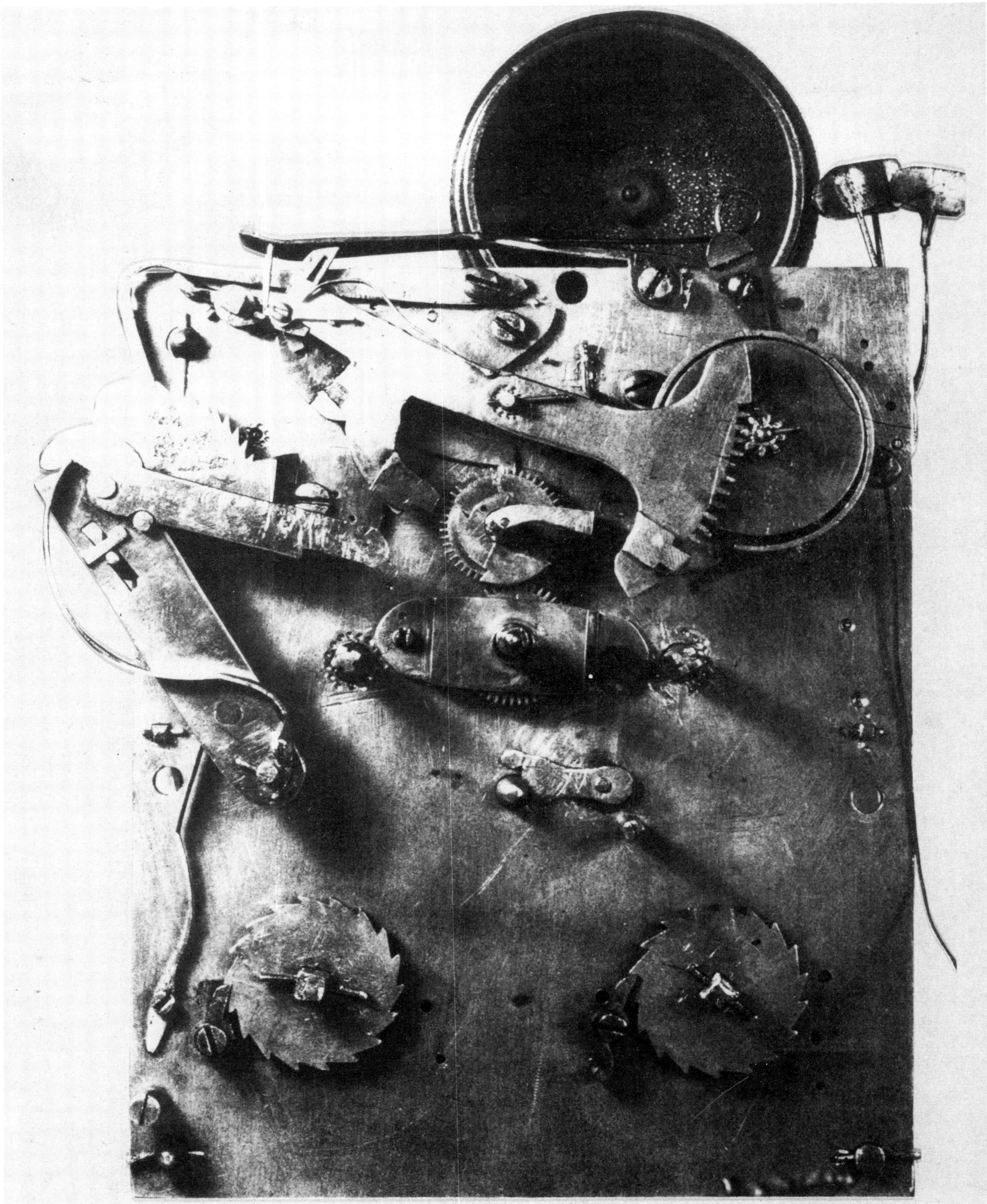

Unidentified. Clock now signed "Tompion". This pull quarter clock was shown by Beresford Hutchinson in his lecture *Fakes and Facsimiles* on 21st May 1982. It has nothing to do with Tompion. Its interest lies in the similarity of its pull quarter to that shown opposite on page 61.

Striking Clock with Pull Quarter on Four Bells

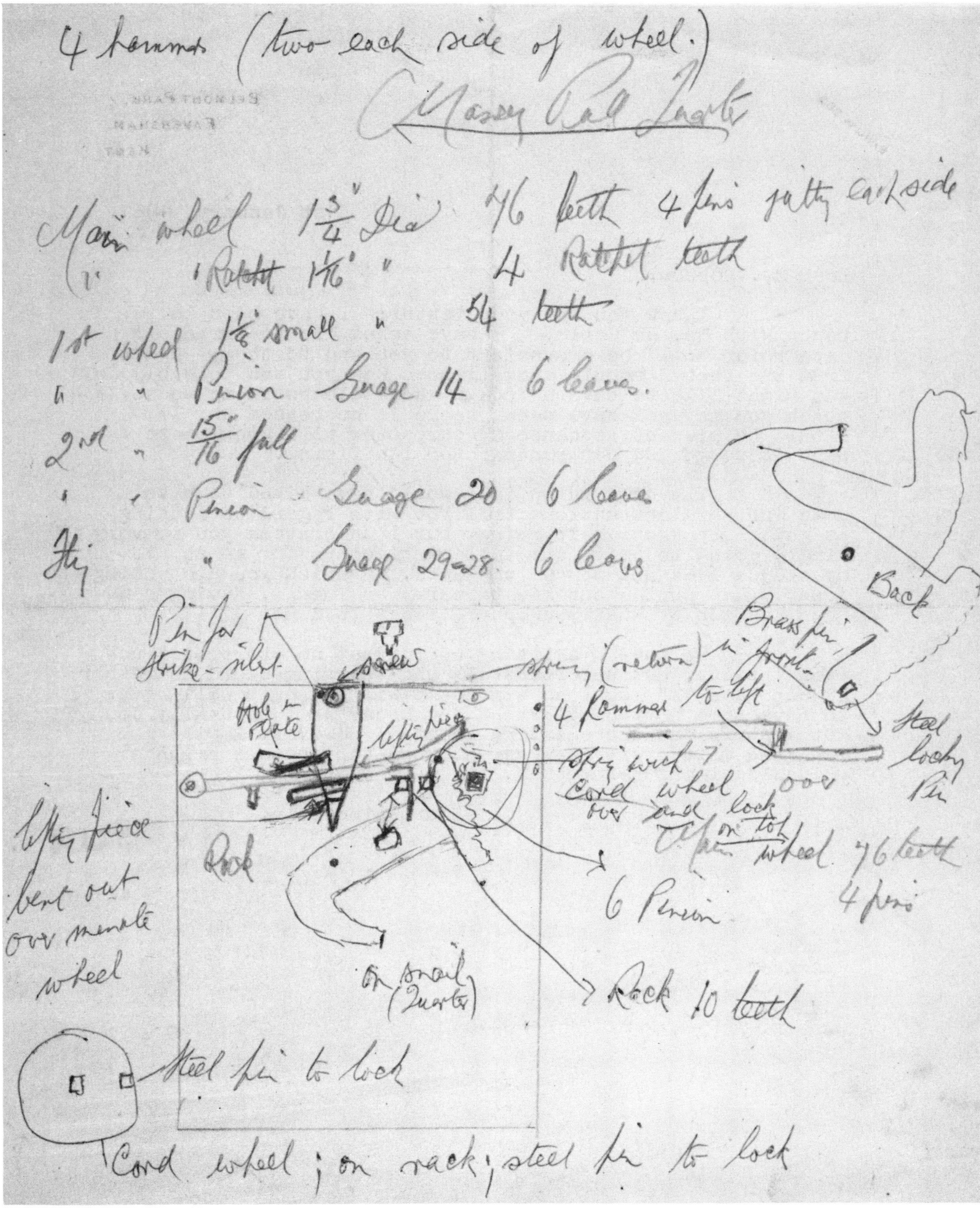

Jacob Massey. View of front plate showing outside quarter rack and pinion and details of pull quarter train.

Striking Clock with Pull Quarter on Three Bells

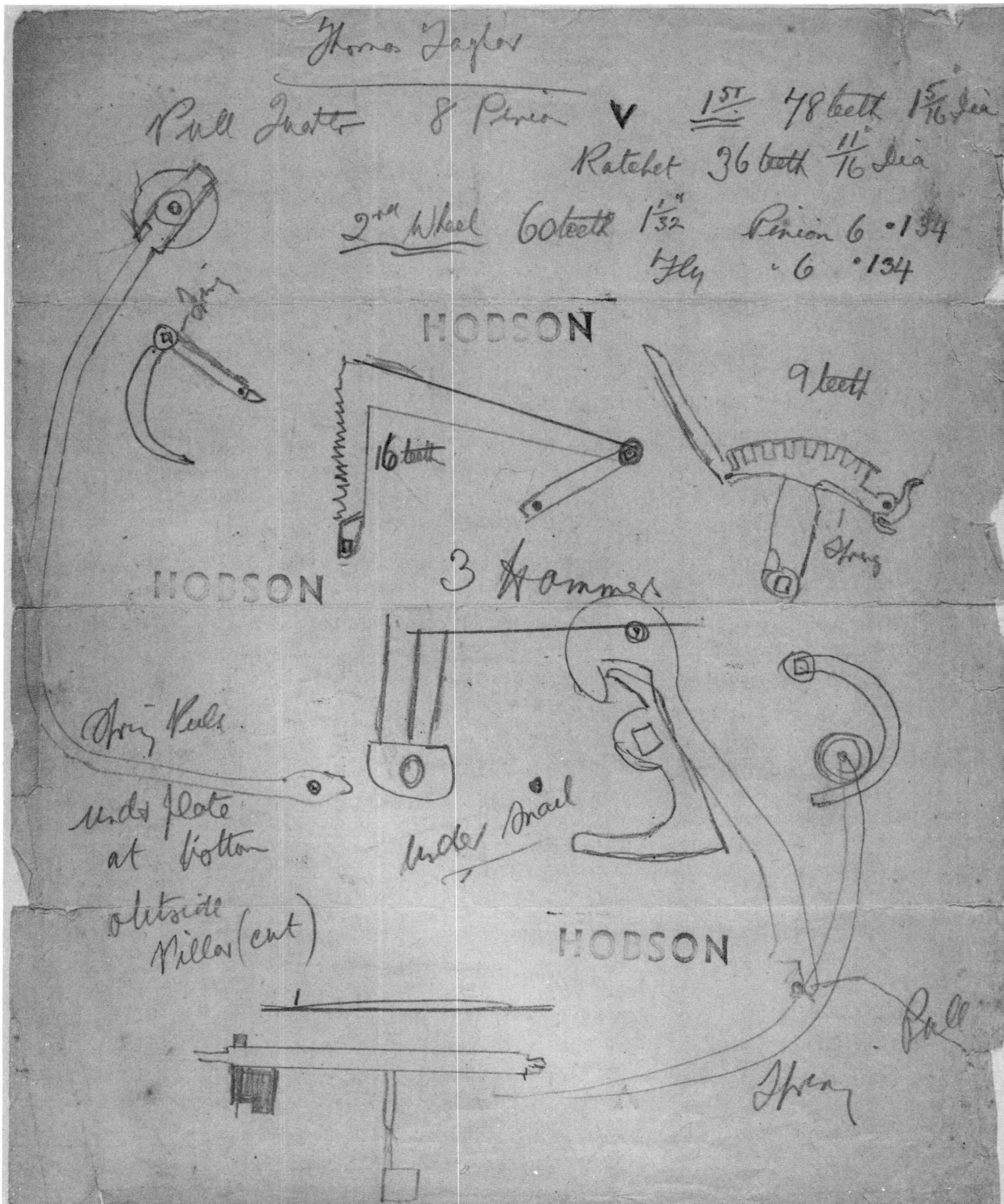

Thomas Taylor (Knibb). Details of pull quarter work and train, internal hour striking rack and two-way pull piece. This clock is almost certainly the work of Joseph Knibb. For further evidence see Lee *Knibb*, plate 135.

Striking Clock with Pull Quarter on Three Bells

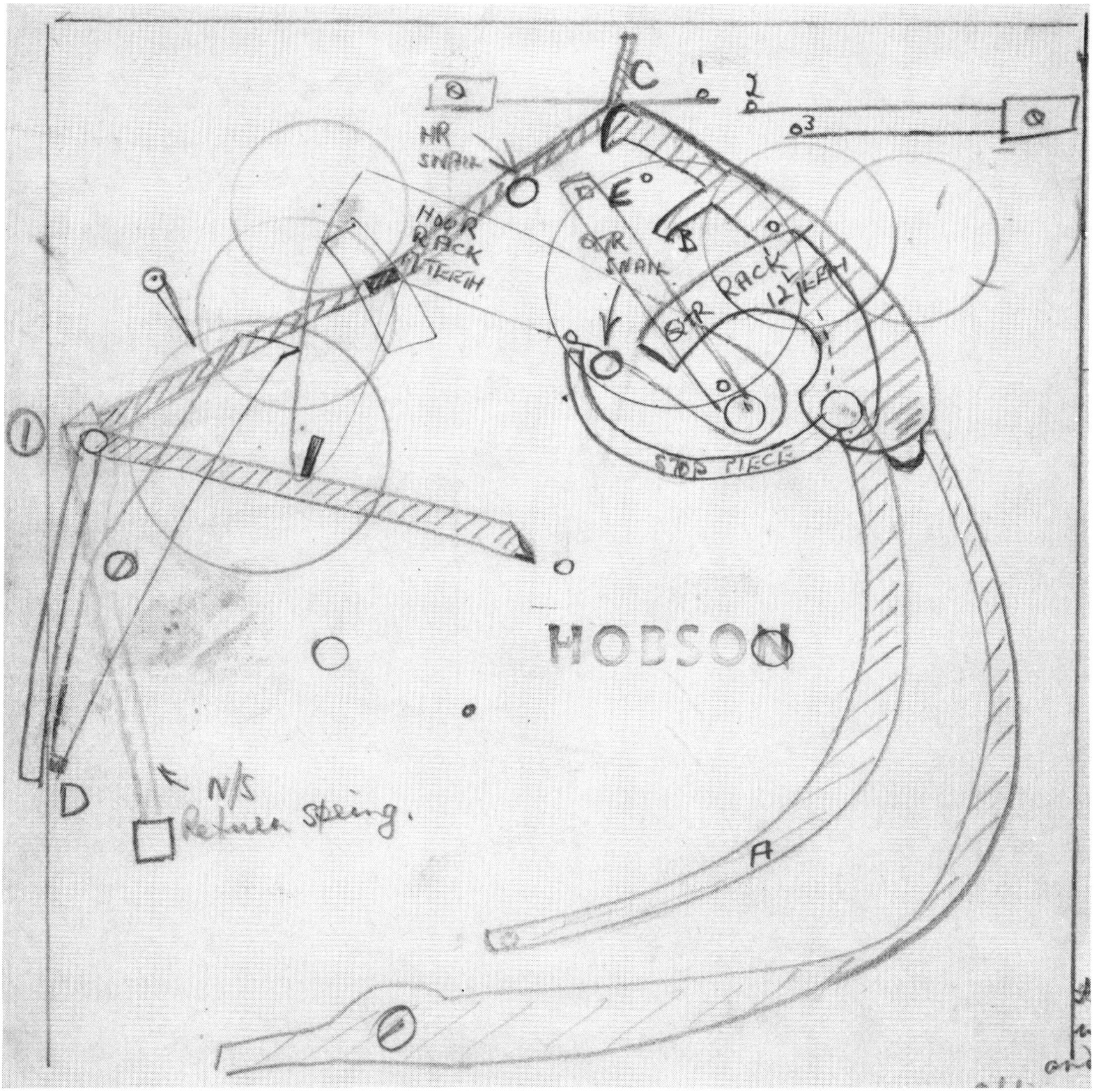

Pull Qtr. on 3 bells by Thomas Taylor.

Striking	Main:	78	*Going*	Main:	72
	Pin:	60-8-10 pins		Centre:	78-6
	Gathering:	48-6		Contrate:	72-6
	Warning:	48-6		Crown:	29-6
Qtr:	Main:	56-8-4 pins			
	Second:	42-6 (in a cock on backplate to qtr. rack)			
	Third:	42-6			

Thomas Taylor. Here Taylor uses a different approach which perhaps owes more to himself than Knibb. On pulling off lever A to the right the quarter rack gathers the main wheel pinion (not shown) until tail B reaches the quarter snail. This action raises lever C. This in turn lifts the rack hook by means of the tails at D, meanwhile holding the striking train on the warn. When the pull is released the heavy blade spring drives the quarter train via the rack causing the quarters to be struck. When the quarter rack has completed its return movement lever C is allowed to fall, thus releasing the striking train.

Timepiece with Silent Pull sounding the Quarters on a Single Bell

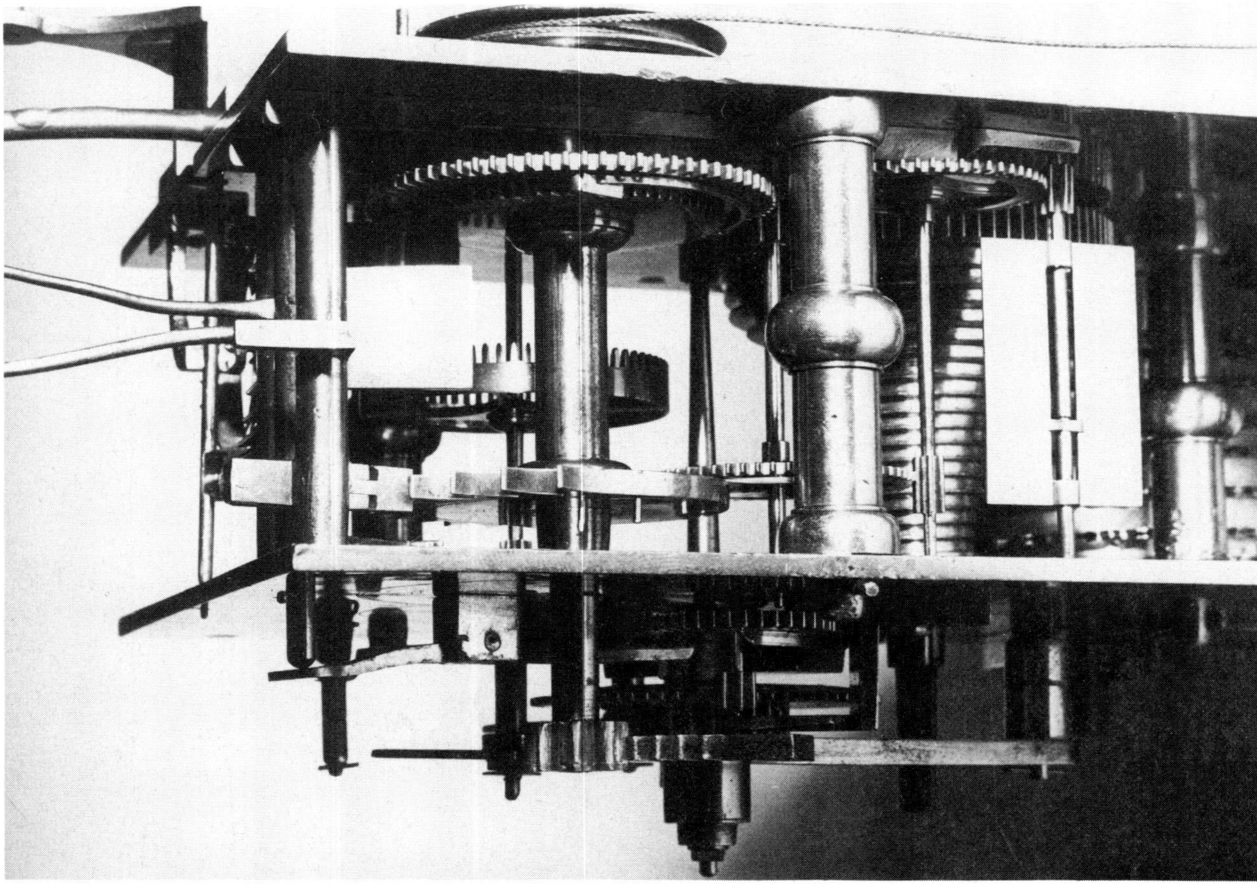

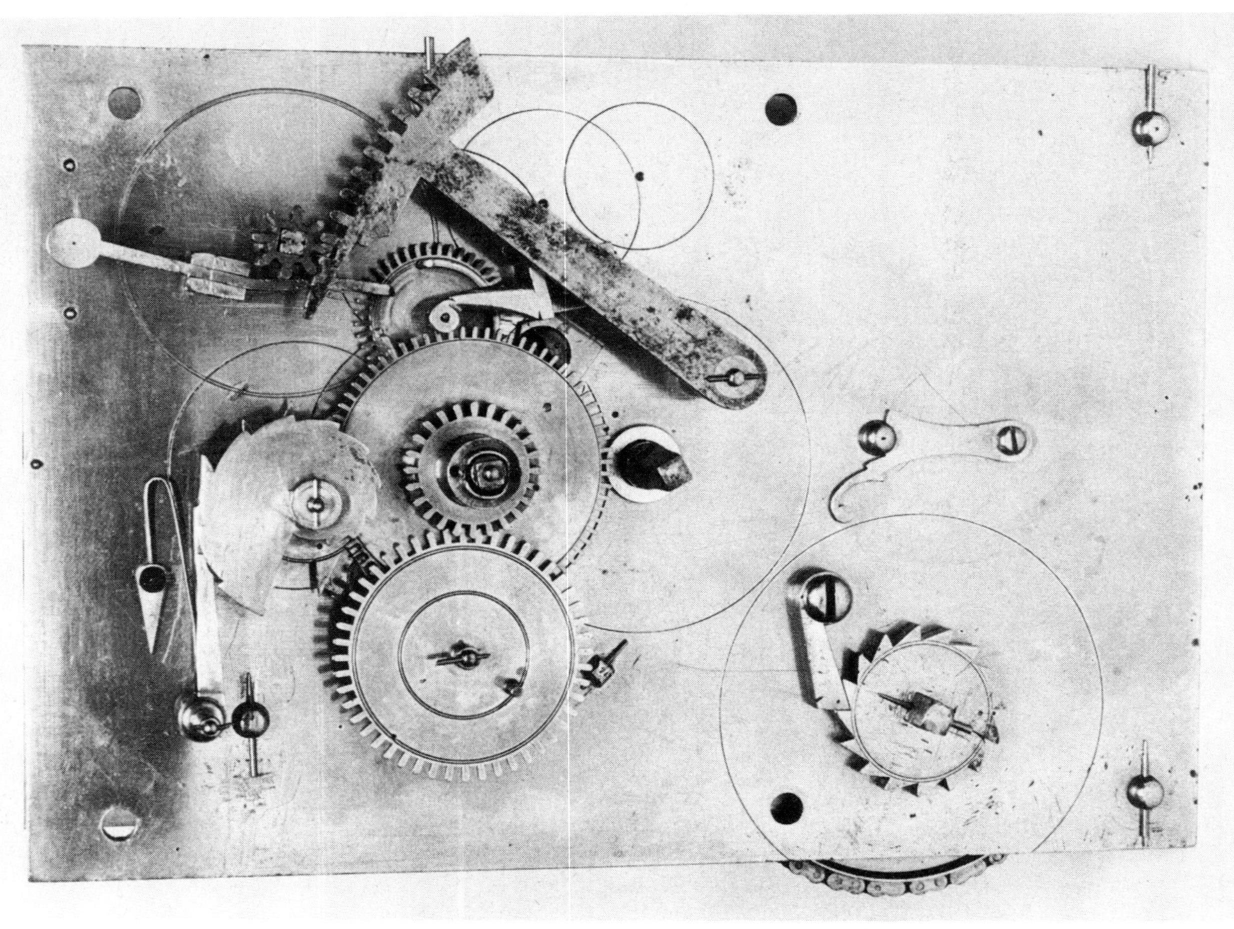

Richard Rayment. See Beauvais opposite on page 65 for sketch of a very similar clock.

Timepiece with Silent Pull sounding the Quarters on a Single Bell and with Alarm

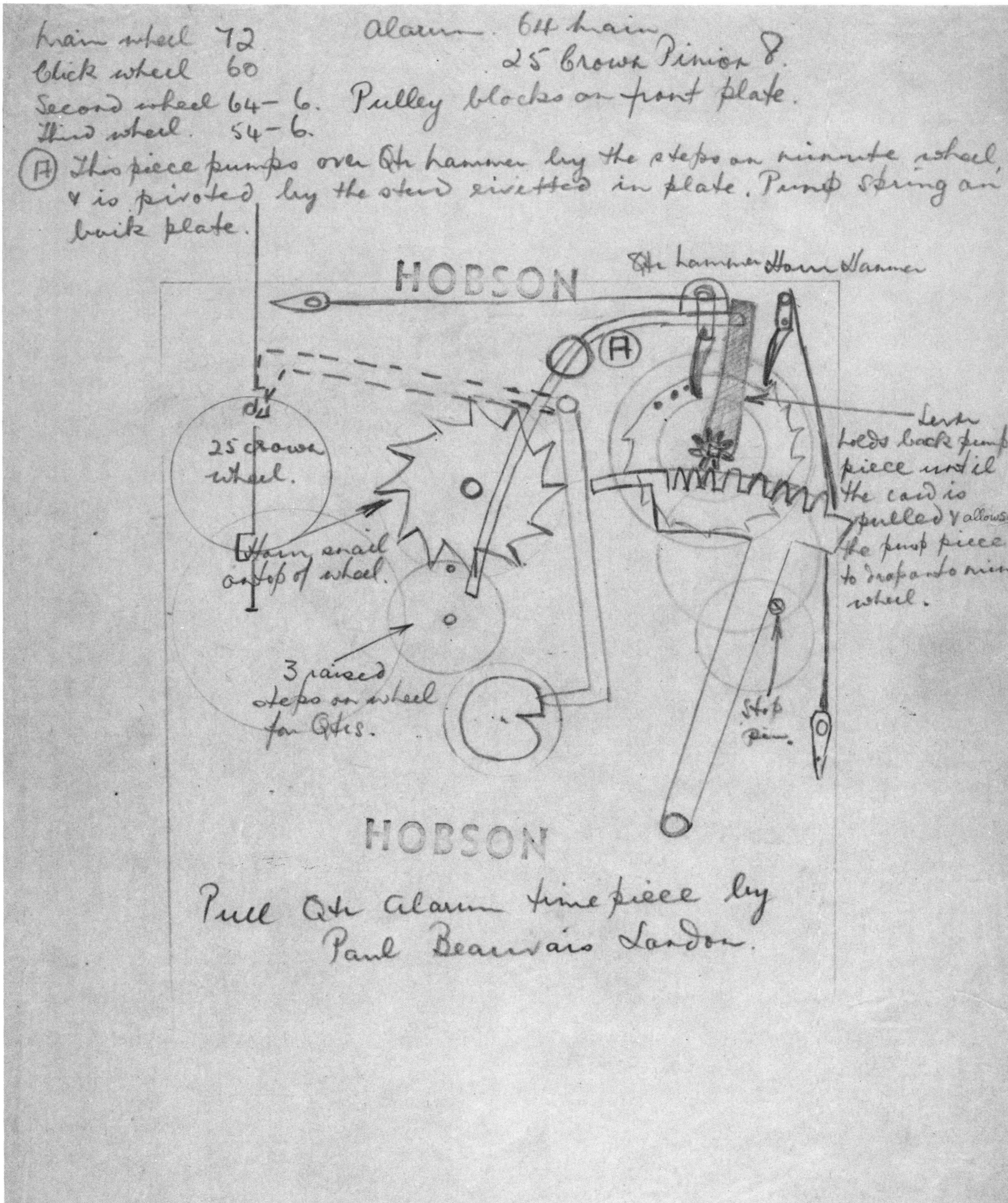

Paul Beauvais. Details of parts showing external rack and pinion and rocking bar. This design is identical in in principle to that of Knibb with the difference that here the hour hammer is lifted by saw-shaped teeth instead of by pins. In this view the reader should imagine that the front plate is transparent.

Timepiece with Silent Pull sounding the Quarters on a Single Bell

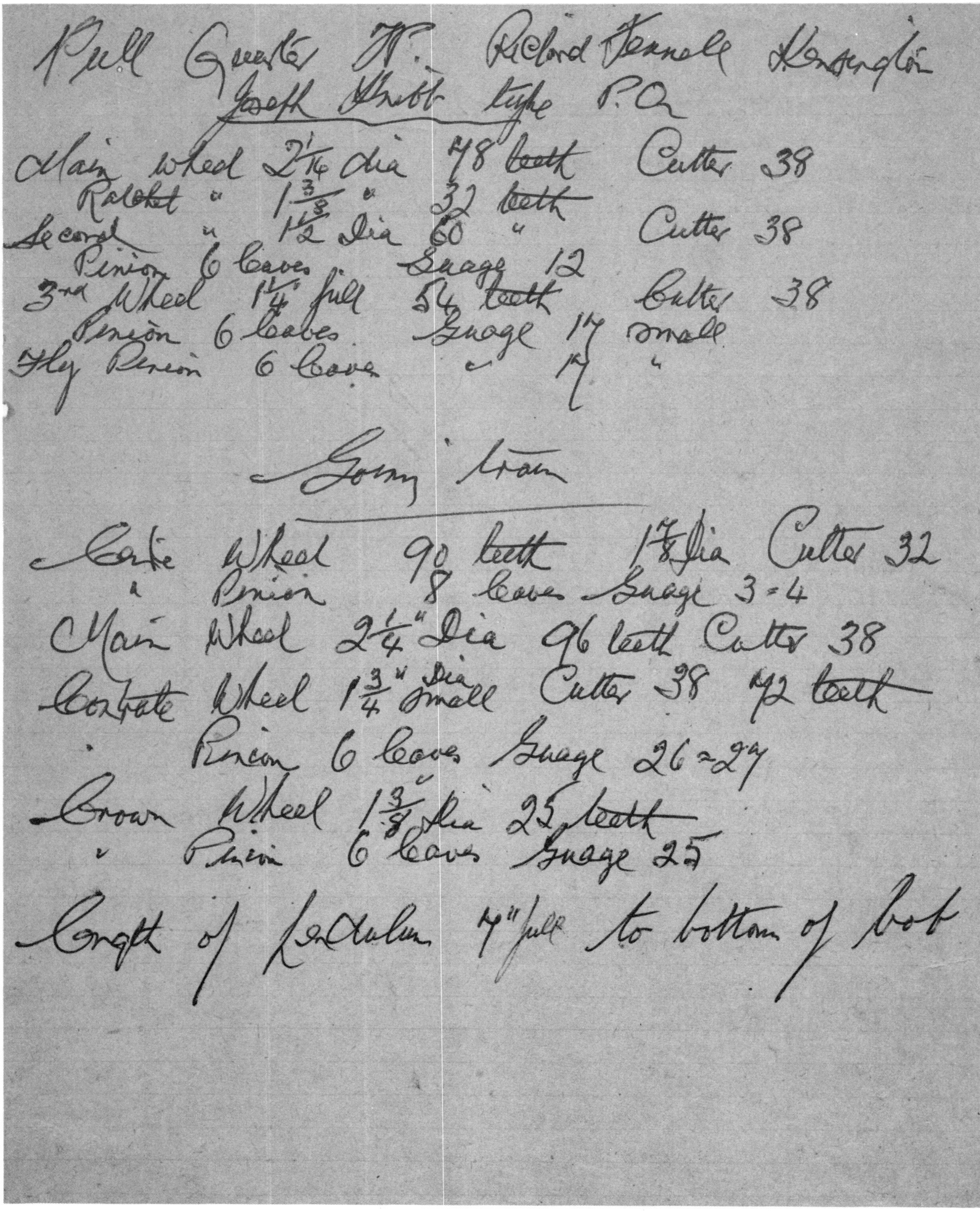

Pull Quarter W. Richard Fennell Kensington
Joseph Knibb type P. Q.
Main Wheel 2 1/16 Dia 78 teeth Cutter 38
Ratchet " 1 3/8 " 32 teeth
Second " 1 1/2 Dia 60 " Cutter 38
Pinion 6 leaves Guage 12
3rd Wheel 1 1/4" full 54 teeth Cutter 38
Pinion 6 leaves Guage 14 small
Fly Pinion 6 leaves " 17 "

Going train

Centre Wheel 90 teeth 1 7/8 Dia Cutter 32
" Pinion 8 leaves Guage 3-4
Main Wheel 2 1/4" Dia 96 teeth Cutter 38
Contrate Wheel 1 3/4" Dia small Cutter 38 72 teeth
Pinion 6 leaves Guage 26-27
Crown Wheel 1 3/8 Dia 25 teeth
" Pinion 6 leaves Guage 25
Length of Pendulum 7" full to bottom of bob

Richard Fennell. Details of wheels and pinions in silent pull and going trains.

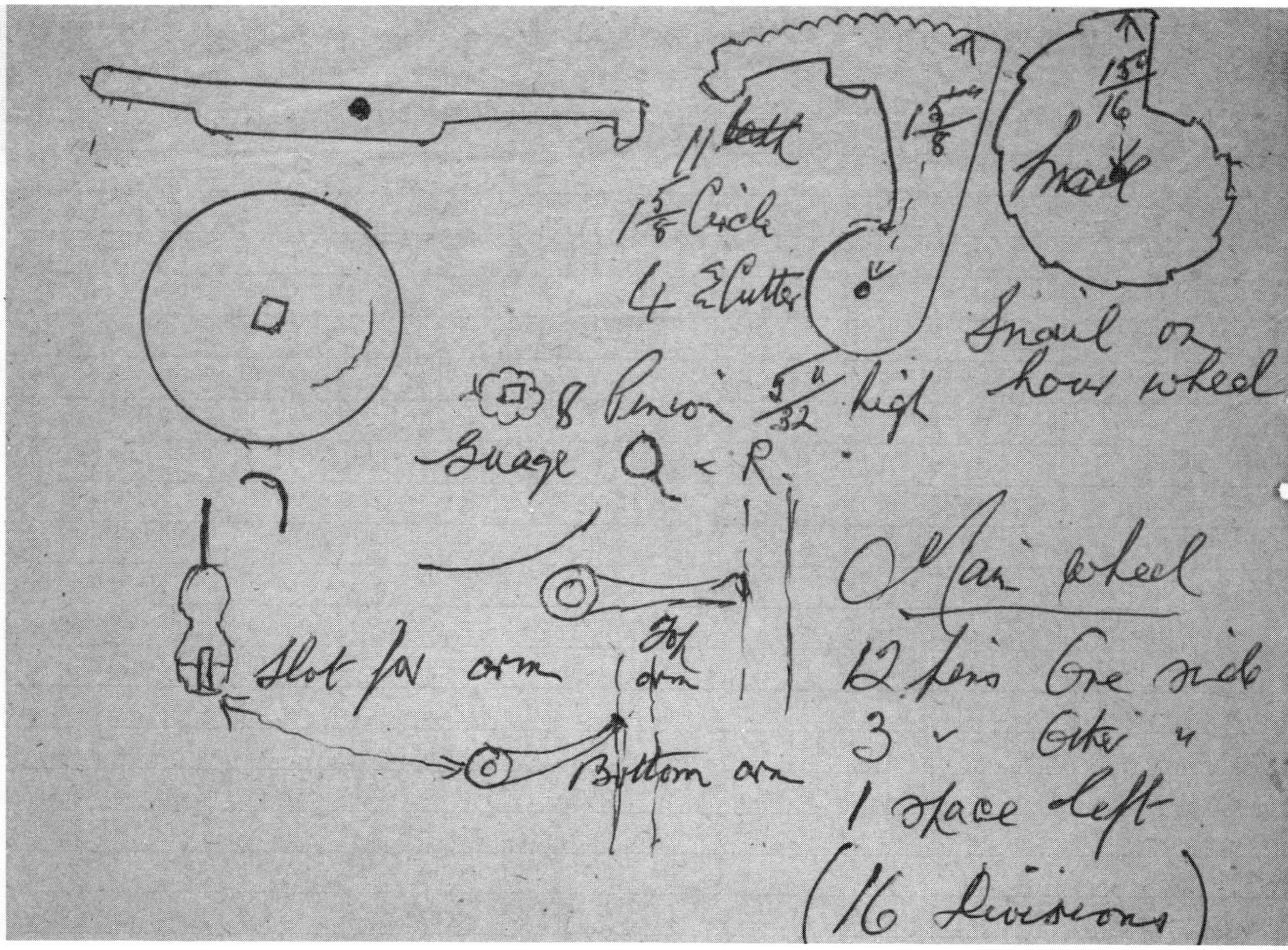

Richard Fennell. Knibb-type silent pull. Sketches of various parts including hammer linkage.

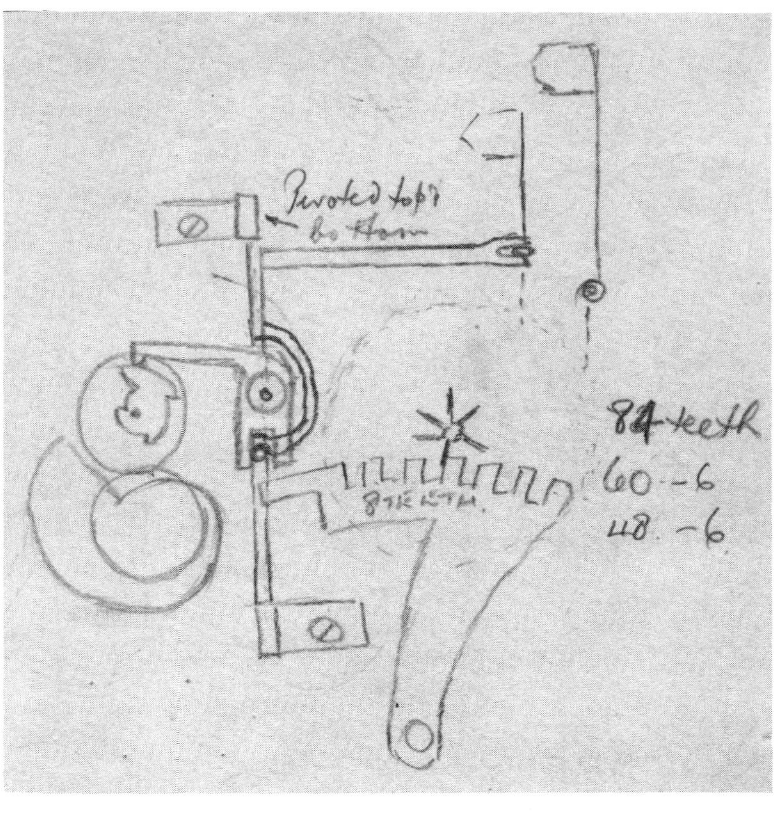

Unidentified maker. Sketch of an early timepiece silent pull system with some similarities to that of Knibb but utilising a slightly more complicated pivoting pump mechanism for the quarters.

ROOKSBY (probably John of York)

Timepiece with Silent Pull sounding the Quarters on a Single Bell

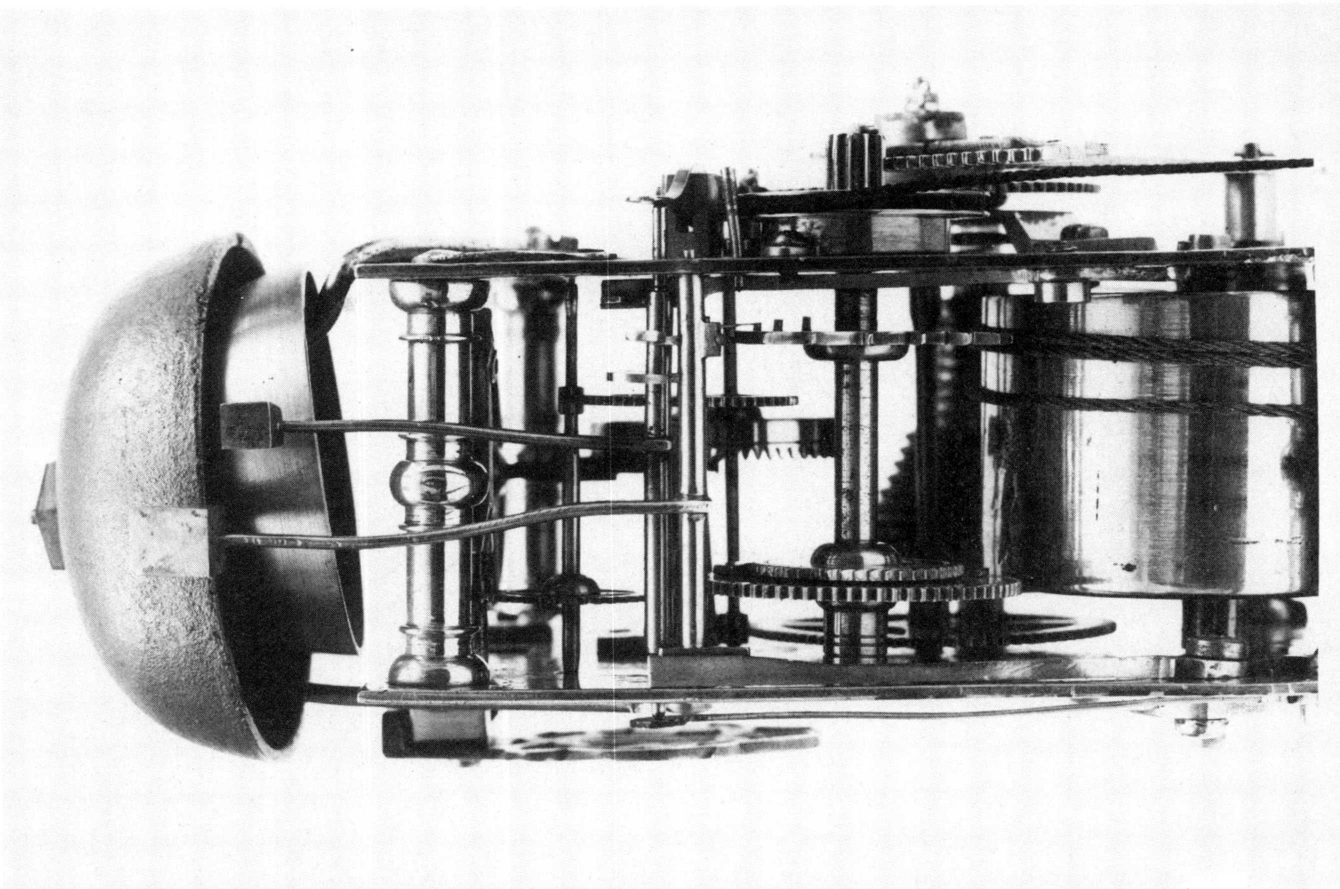

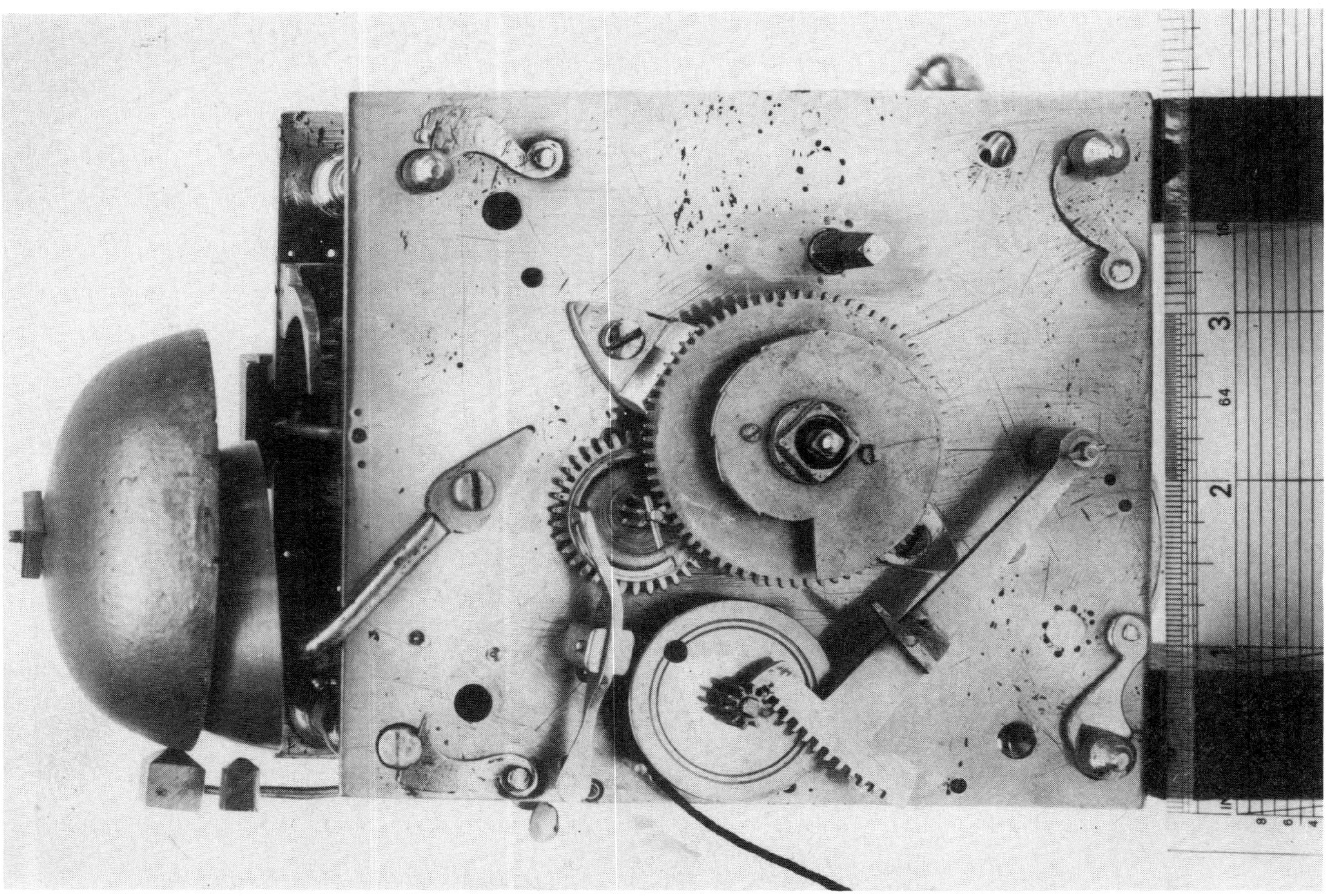

Rooksby. A particularly small provincial Knibb-type silent pull.

Timepiece with Silent Pull sounding the Quarters on Three Bells

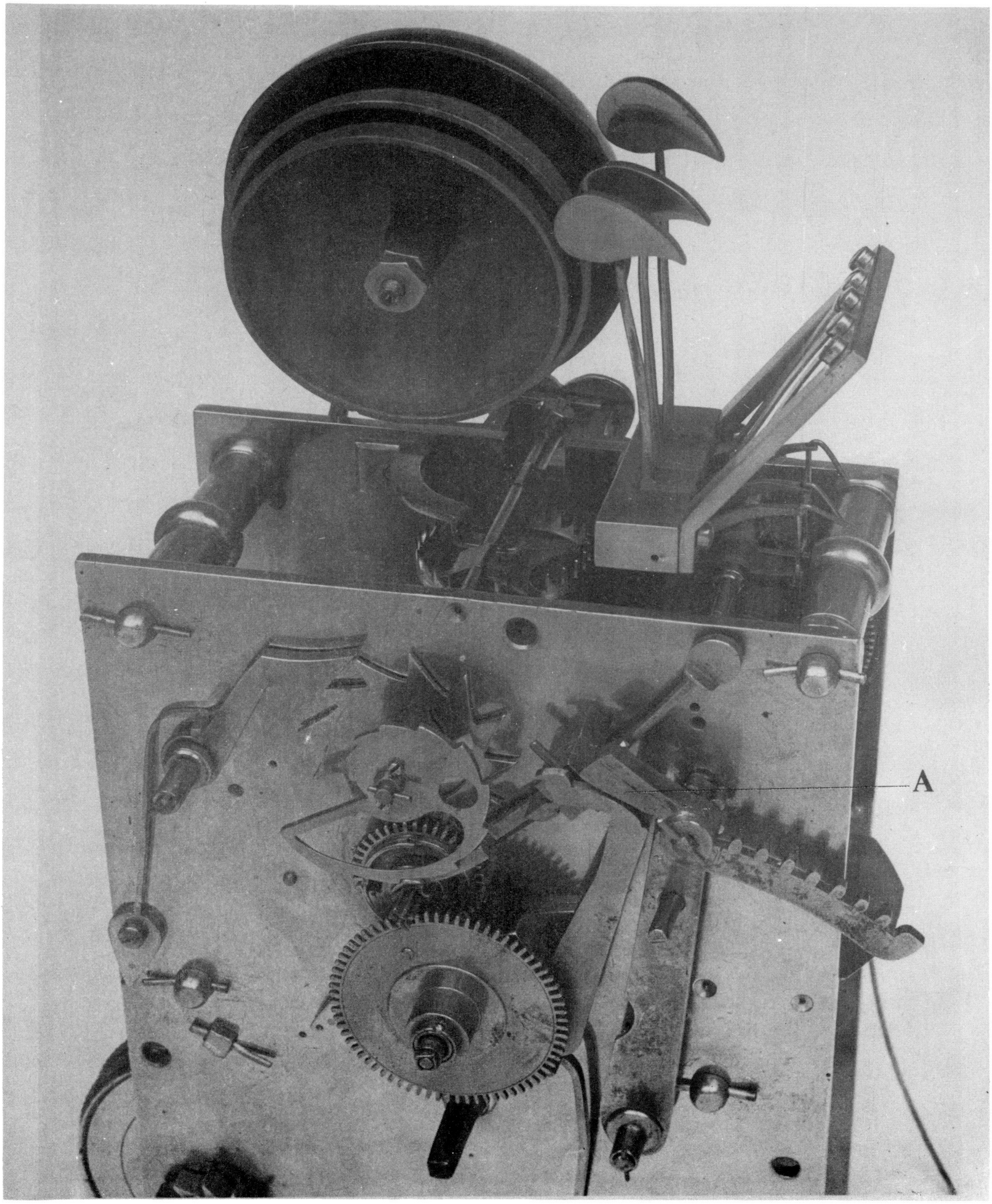

Ellicott. In this example the rocking bar is held away from the quarter snail by the bevelled lever A until the pull off.

Striking clock with Pull Quarter on a Single Bell

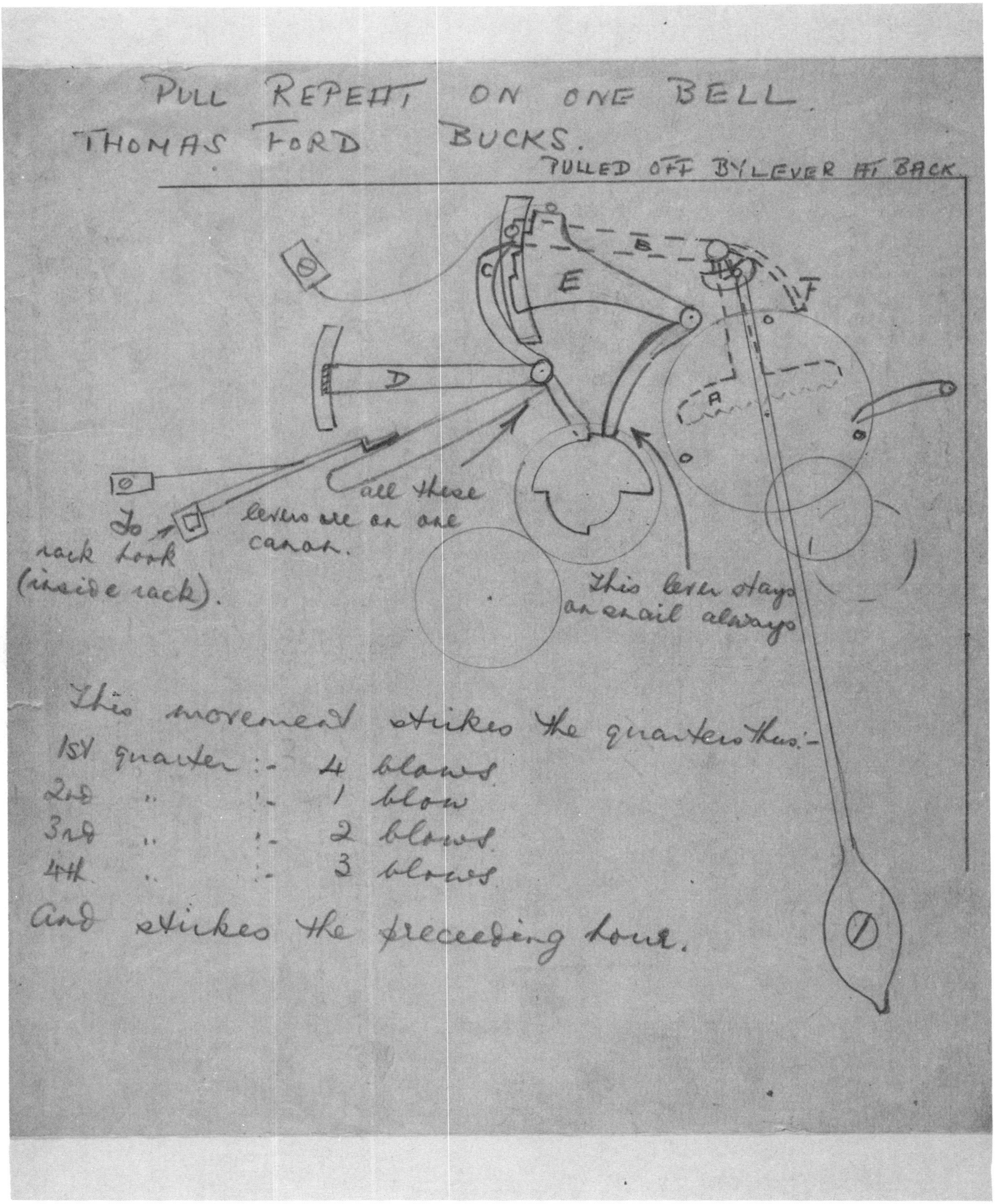

Thomas Ford. Details of pull-quarter and striking work. Note the unusual stepped quarter rack. Date probably post 1724; but this design shows many early features.

Striking Clock with Pull Quarter on a Single Bell

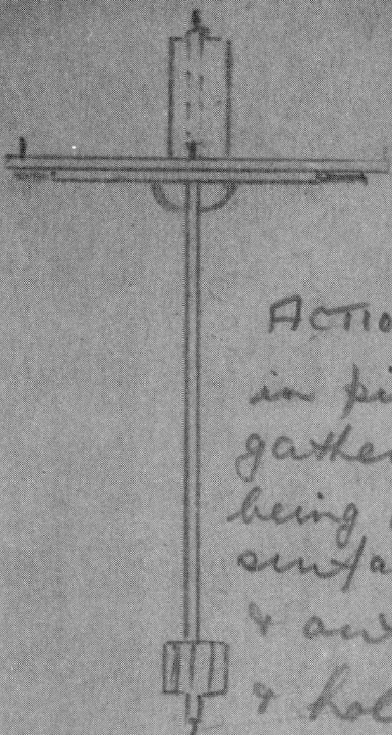

MAIN WHEEL:- 60 TEETH. 3 PINS.
CLICK. WHEEL:- 15 TEETH.
PINION :- 6.
TEETH ON PULLING OFF BAR:- 11.
SECOND WHEEL: 48 TEETH 6 PINION

ACTION:- On pulling off, lever A (engaging in pinion) pulls round main arbor & gathers up the click wheel. Lever B being on the same arbor as A bears on the surface of lever C, pushing it downwards & outwards so lifting the rack hook & holding the striking by the warning piece 'D'. Lever B is allowed to travel as far as a step on lever E will allow this being determined by the position of snail. On releasing lever, the return spring pushes detent with levers A & B attached allowing quarters to be rung struck. Levers C & D being spring loaded return to their normal position allowing the hours to be struck. Lever F being on the same detent as A & B stops the quarters fly jamming in the teeth of the main wheel.

Thomas Ford. Pull quarter train and its action. The main wheel and clickwork are shown top left.

WINDMILLS
(probably Joseph)

Striking Clock with Pull Quarter on Four Bells

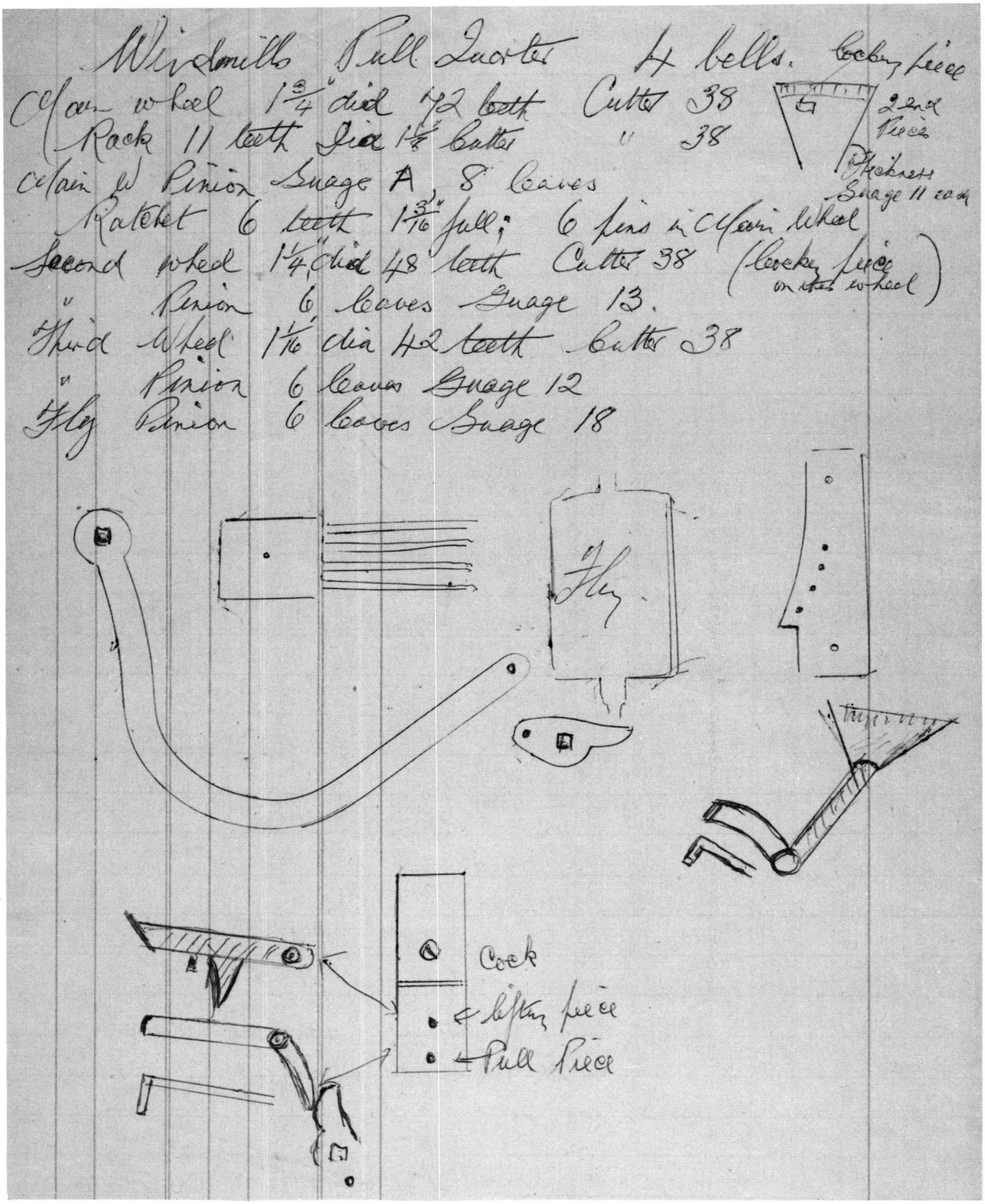
Windmills Pull Quarter 4 bells.

Main wheel $1\frac{3}{4}$" dia 72 teeth Cutter 38

Rack 11 teeth Dia $1\frac{1}{8}$" Cutter " 38

Main W Pinion Guage A, 8 leaves

Ratchet 6 teeth $1\frac{3}{16}$" full; 6 pins in Main Wheel

Second wheel $1\frac{1}{4}$" dia 48 teeth Cutter 38 (locking piece on this wheel)

" Pinion 6 leaves Guage 13.

Third Wheel $1\frac{1}{16}$ dia 42 teeth Cutter 38

" Pinion 6 leaves Guage 12

Fly Pinion 6 leaves Guage 18

Windmills. Details of pull quarter train and of some parts including inside quarter rack, hammer spring block, pull off lever and release work for hour striking. Windmills father and son favoured pull quarter work using four, five or six bells.

Striking Clock with Pull Quarter on Three Bells

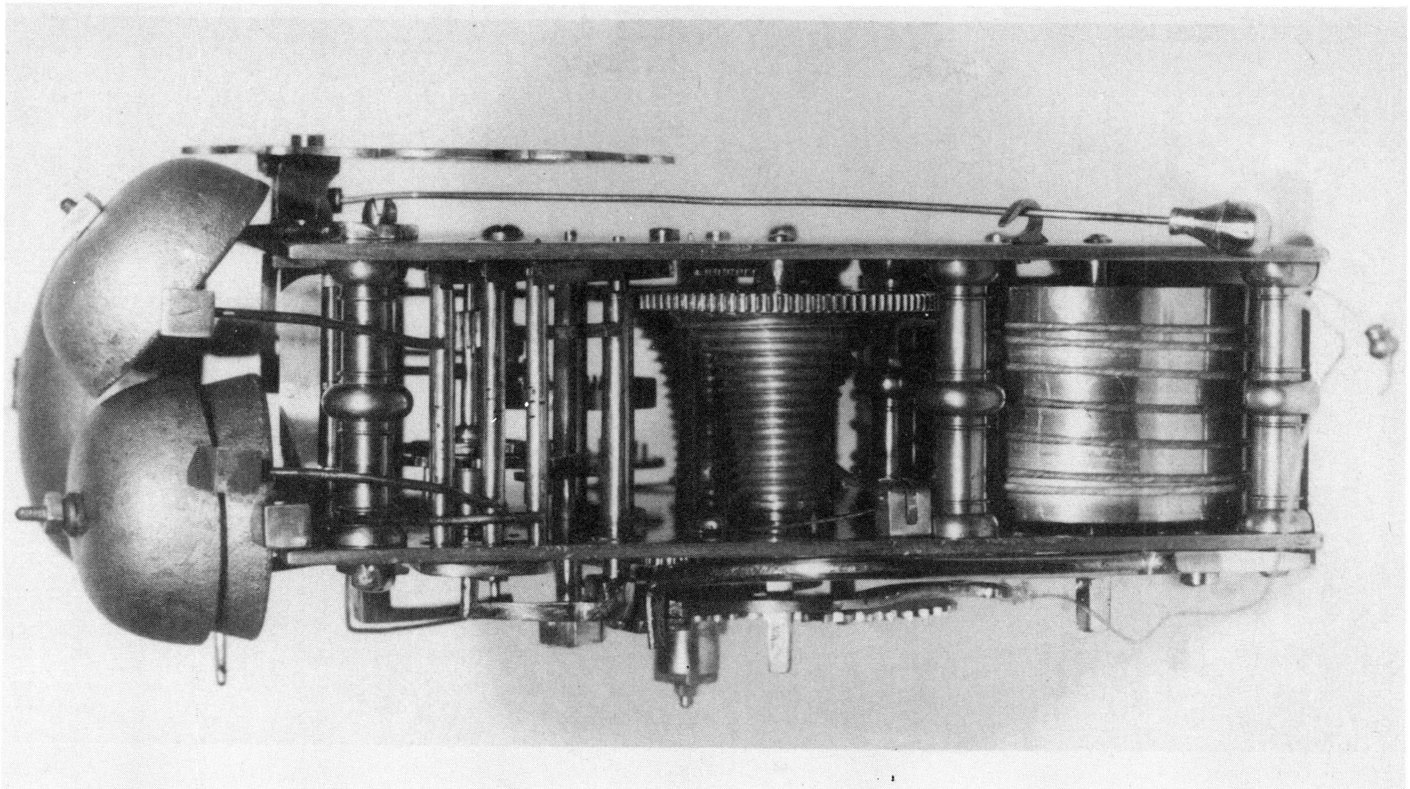

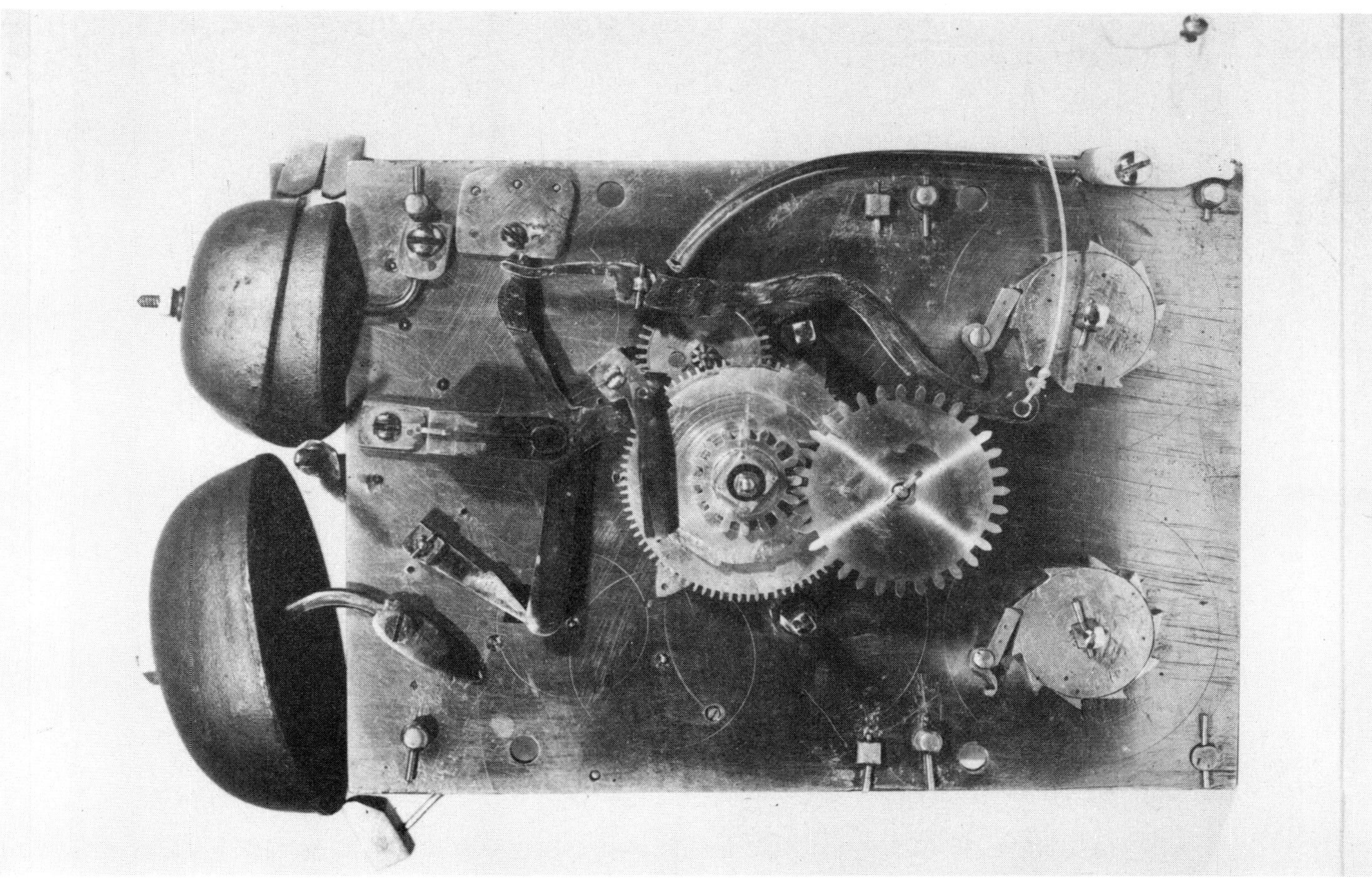

Richard Baker. This clock bears some resemblance to that by Windmills shown on page 72.

Striking Clock with Pull Quarter on Five Bells

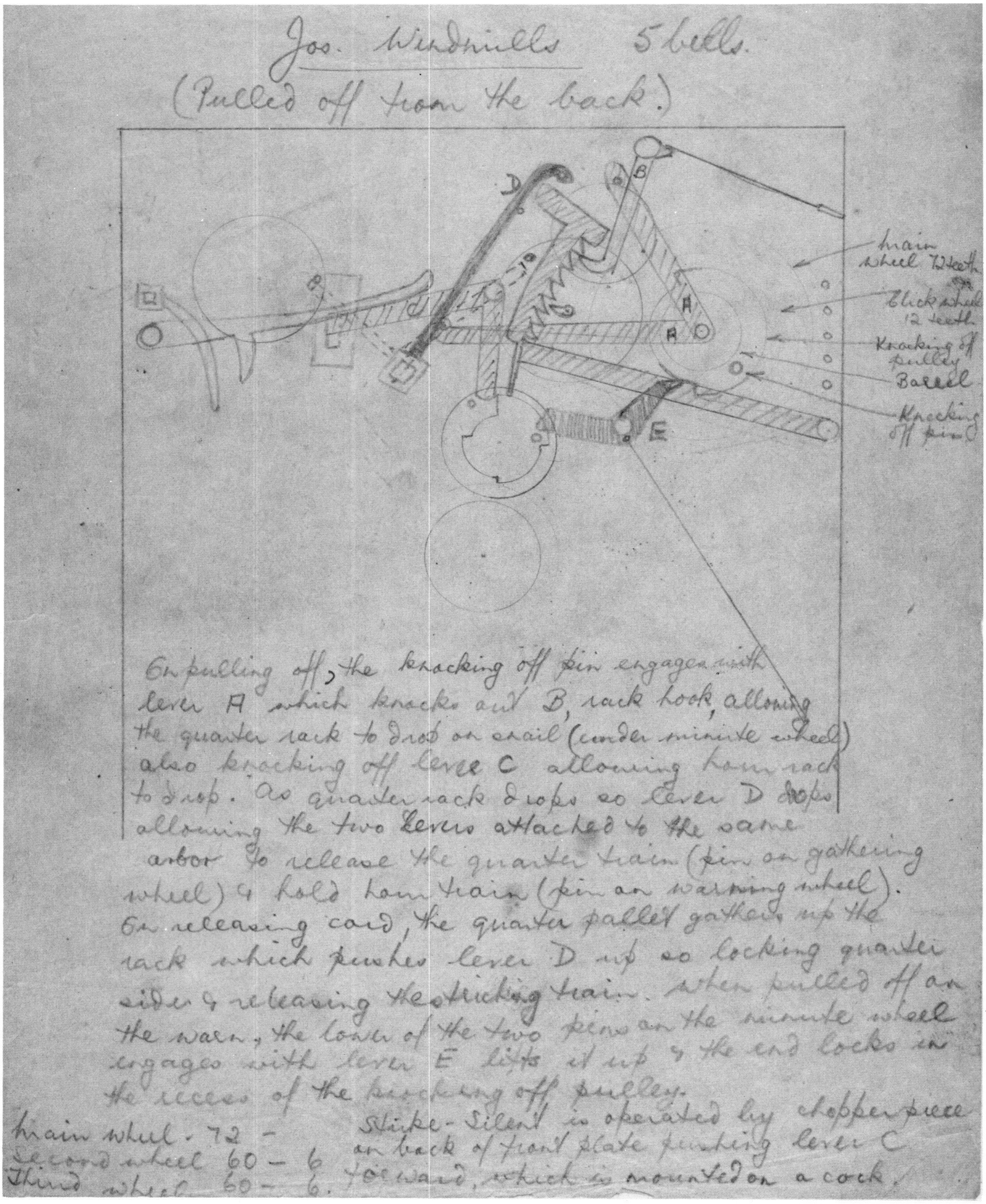

Joseph Windmills. Sketch showing early gathered quarter rack (as opposed to rack and pinion) with train counts and description of action.

Striking Clock with Pull Quarter on Five Bells

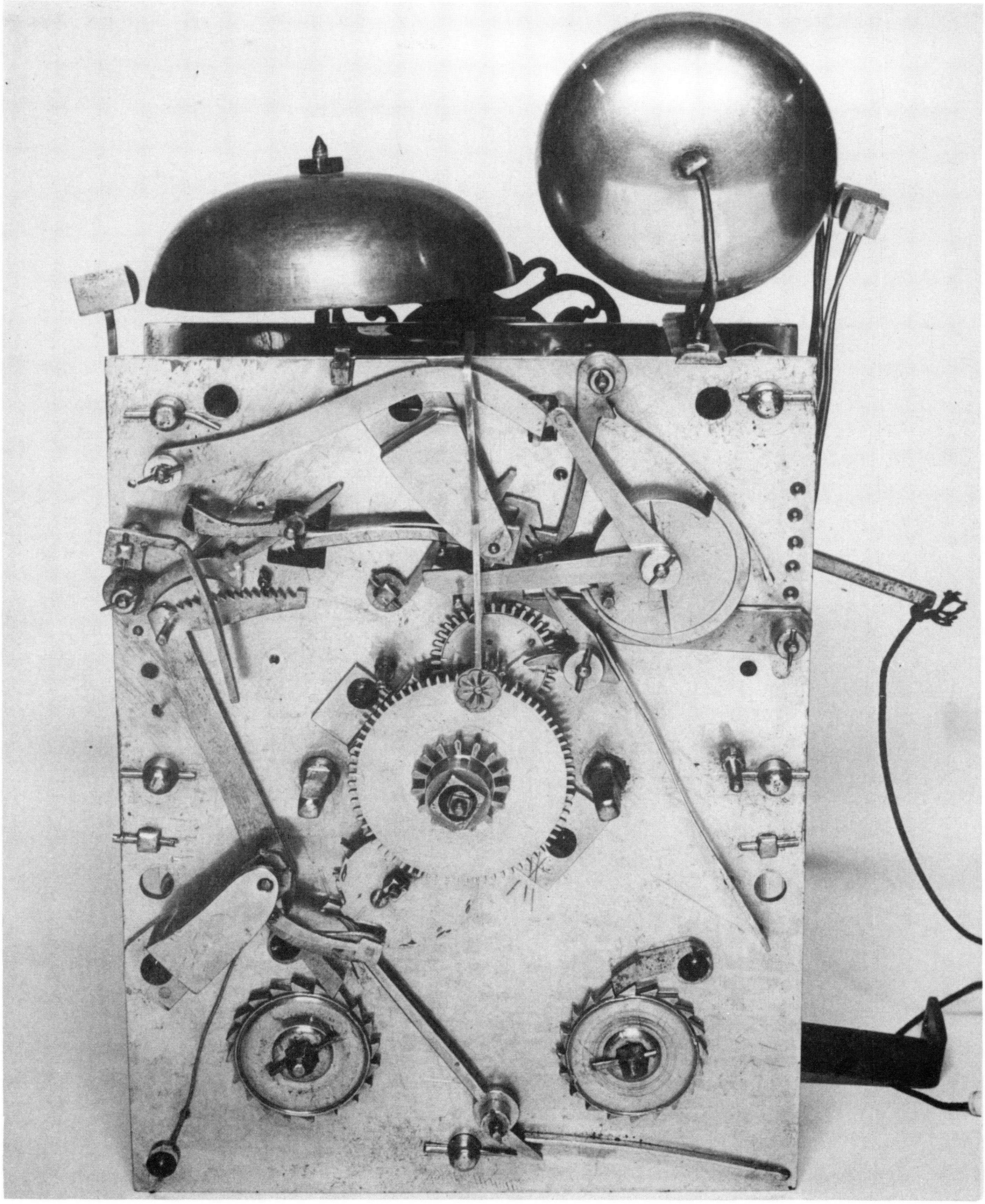

Daniel Quare. This clock (shown pulled off) although signed by Daniel Quare is more reminiscent of Windmills.

Timepiece with Silent Pull sounding the Quarters on Two Bells

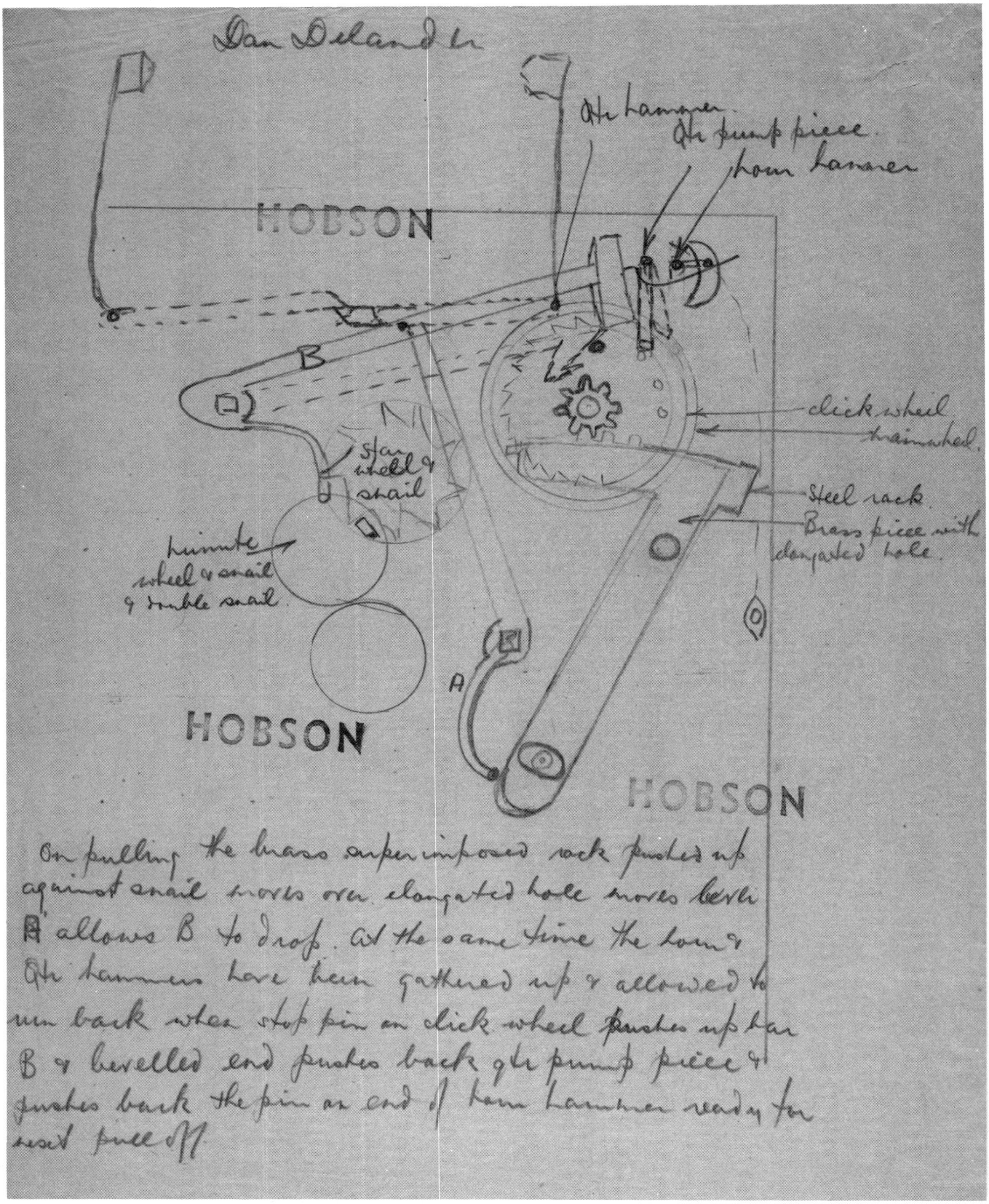

Daniel Delander. Silent pull based on Knibb (see Rayment, pages 64 and 65; also see page 80). Note all-or-nothing is achieved by the superimposed lever on the motion rack. The bevelled block with cranked pin serves to keep the hour and quarter hammer tails clear of the pin wheel until required. See similar clock on pages 78 and 79.

Timepiece with Silent Pull sounding the Quarters on Two Bells

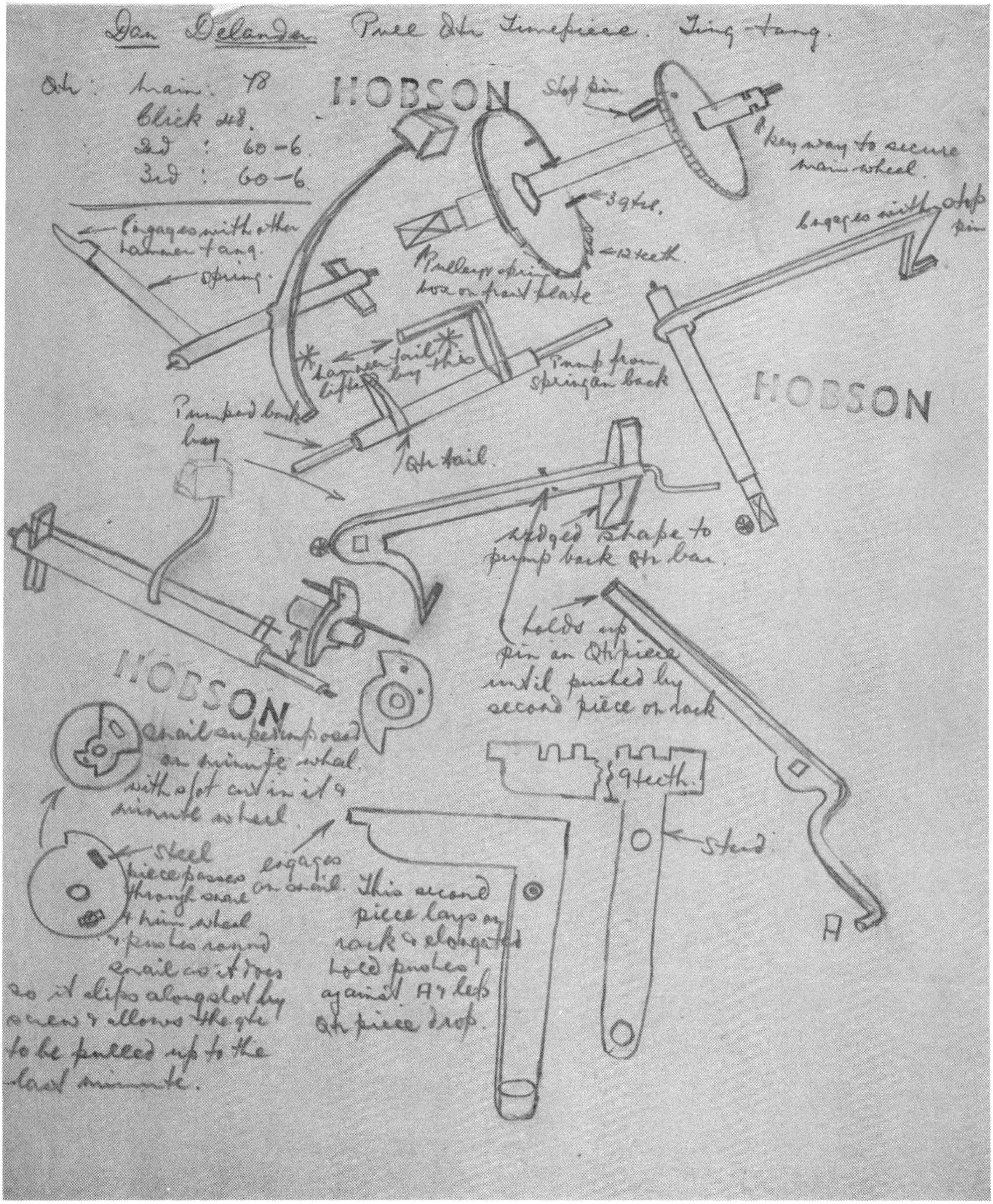

Daniel Delander. Further details of work shown on page 76.

Timepiece with Silent Pull sounding the Quarters on a Single Bell

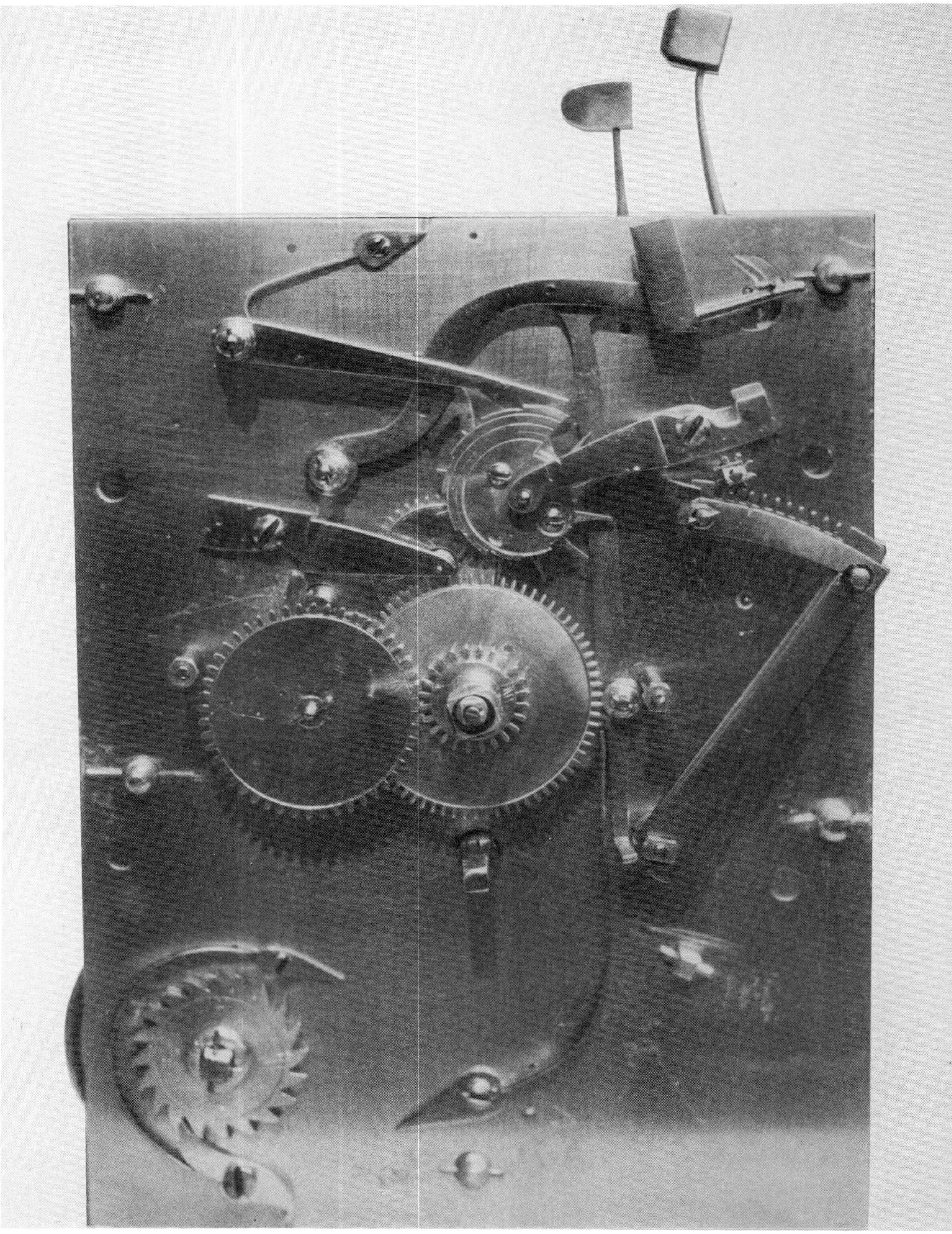

Francis Gregg. This silent pull is similar to those shown on pages 76,77 and 80.

Timepiece with Silent Pull sounding the Quarters on a Single Bell

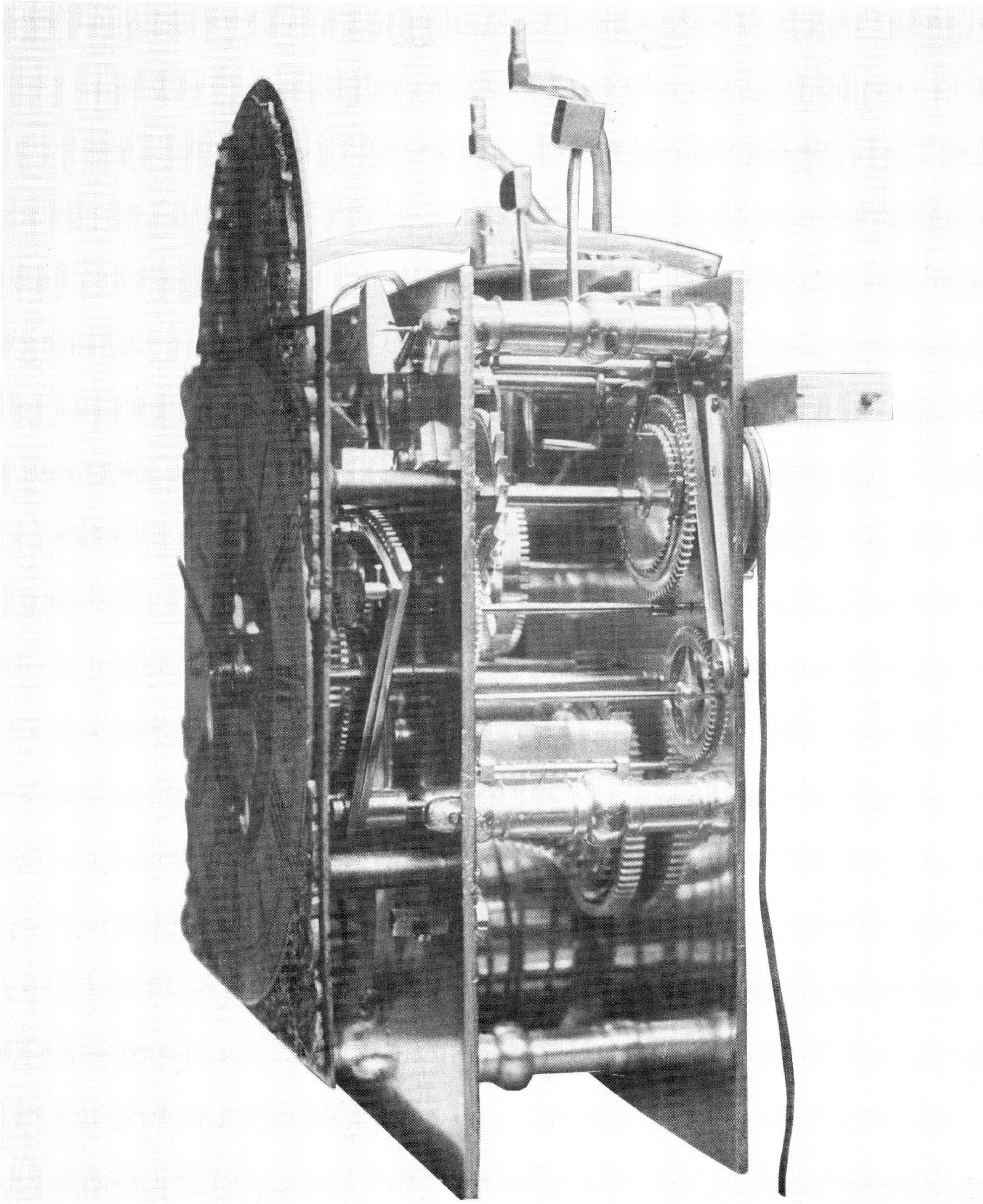

Francis Gregg. Side view of silent pull shown on page 78.

UNIDENTIFIED
(probably
Delander)

Timepiece with Silent Pull sounding the Quarters on Six Bells

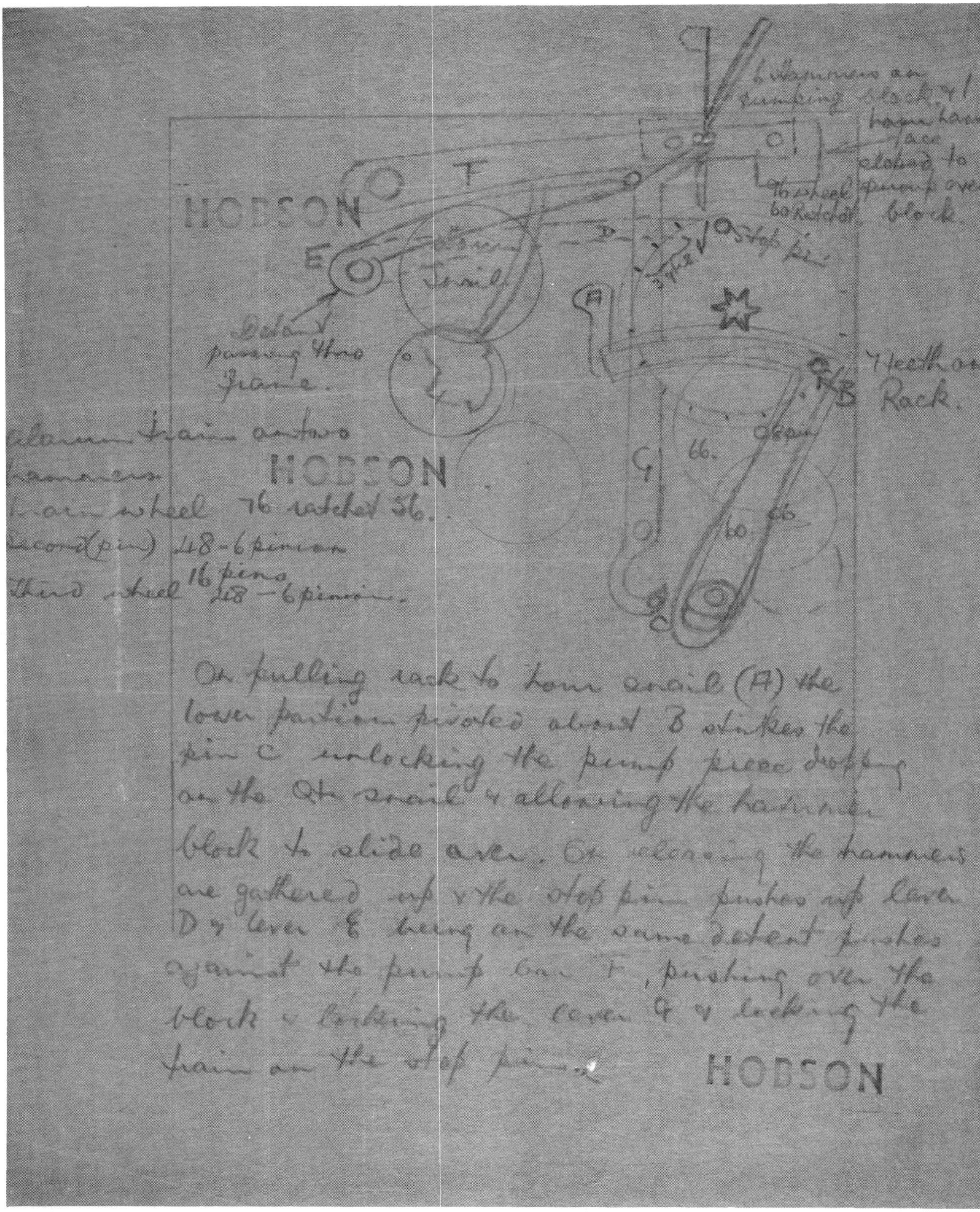

Unidentified. This system is similar in design and operation to those shown on pages 76-79. Note that here a pumped hammer block is used.

Striking Clock with Pull Quarter on a Single Bell and with Alarm

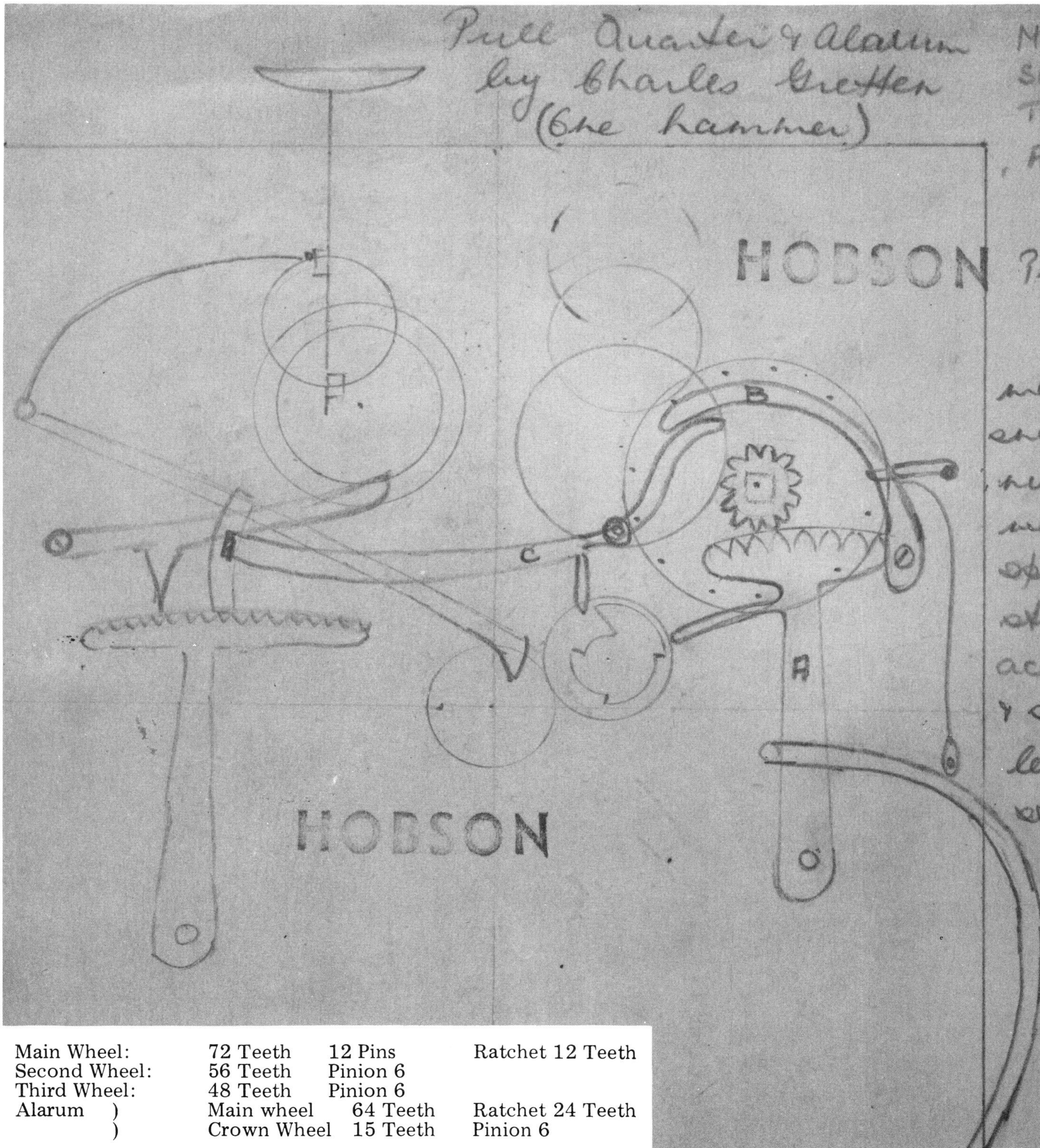

Main Wheel:	72 Teeth	12 Pins	Ratchet 12 Teeth
Second Wheel:	56 Teeth	Pinion 6	
Third Wheel:	48 Teeth	Pinion 6	
Alarum)	Main wheel	64 Teeth	Ratchet 24 Teeth
)	Crown Wheel	15 Teeth	Pinion 6

On pulling off, the pinion on the main wheel drives the outside quarter rack A until checked by the quarter snail gathering up the requisite number of teeth on the click wheel and cocking the curved motionspring. At the same time lever B under pressure of its spring pushes down tail of lever C thereby warning the hour striking train. When the pull is released the motionspring restores the rack causing the quarters to be sounded. On its full return the rack lifts lever B so as to release the hour striking work.

Charles Gretton. Sketch of unusual pull quarter work with train count and description of action.

Striking Clock with Pull Quarter on Four Bells

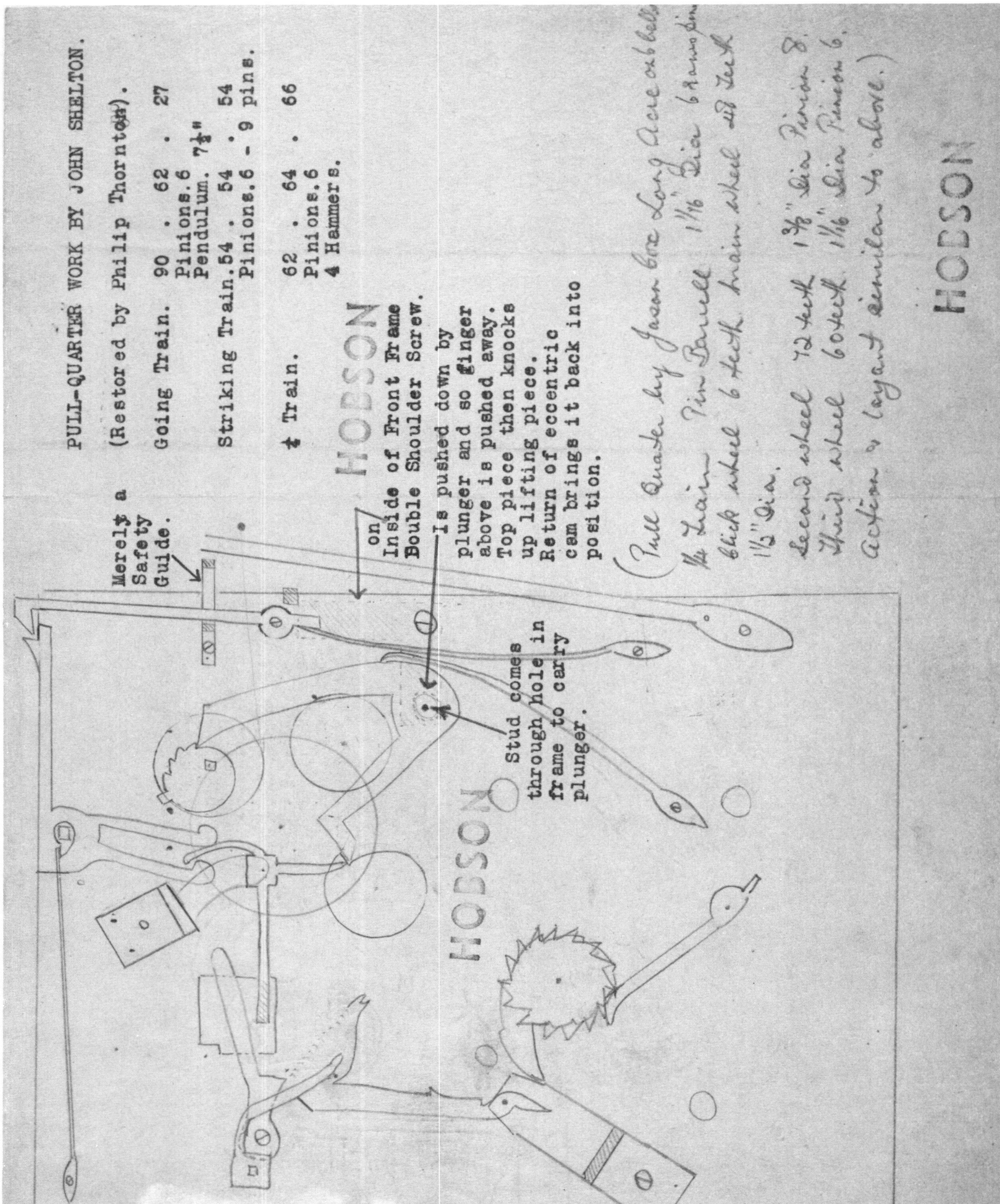

John Shelton. This system, shown here in the pulled state, is very often found used with a pin barrel rather than with a pin wheel. The most unusual vertical lever which releases the hour striking train may have been designed in recent times by the late Philip Thornton. See page 83 for description of action in a smilar clock.

Striking Clock with Pull Quarter on a Single Bell

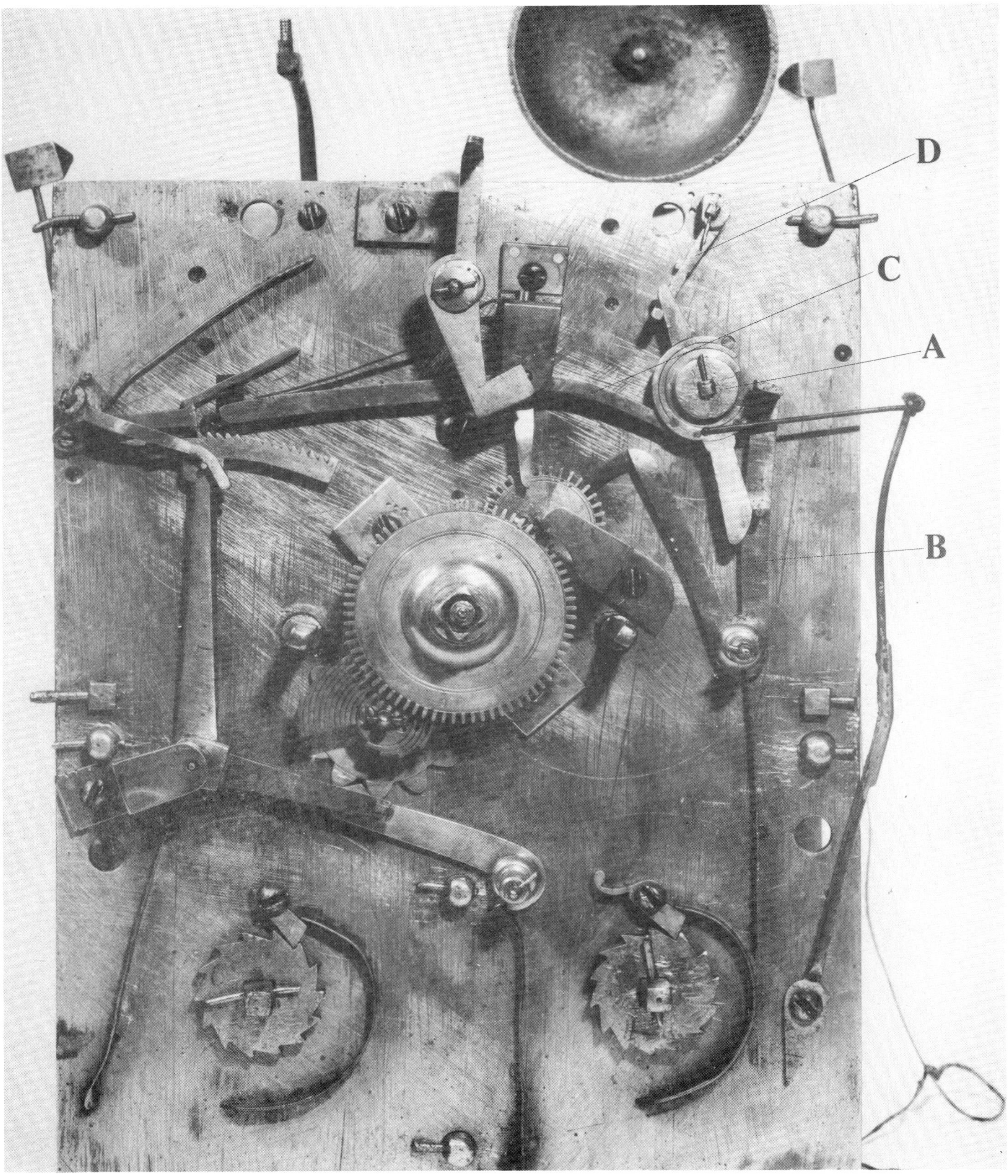

Gould. When the cord is pulled, the stepped wheel A turns clockwise and allows the "V"-shaped double lever to sense the quarter snail. The snail thus controls the angular rotation of A so determining the quarters. At the same time the hour striking train has been warned through the depressing of the lifting lever tail C. See pages 86 and 88 for sketches showing slot arrangement for obtaining the necessary downward motion of C when wheel A is turning clockwise. The wing on the top of A is for locking the pull quarter train by means of a lever D which blocks the path of a pin carried on the second wheel of the runner train. See page 99 for view of the back plate and pull off lever. Thiout showed very much this arrangement in 1741, attributing it to Massy.

NICHOLAS MASSY A Rare Month Striking Clock with Silent Pull Sounding the Quarters on a Single Bell

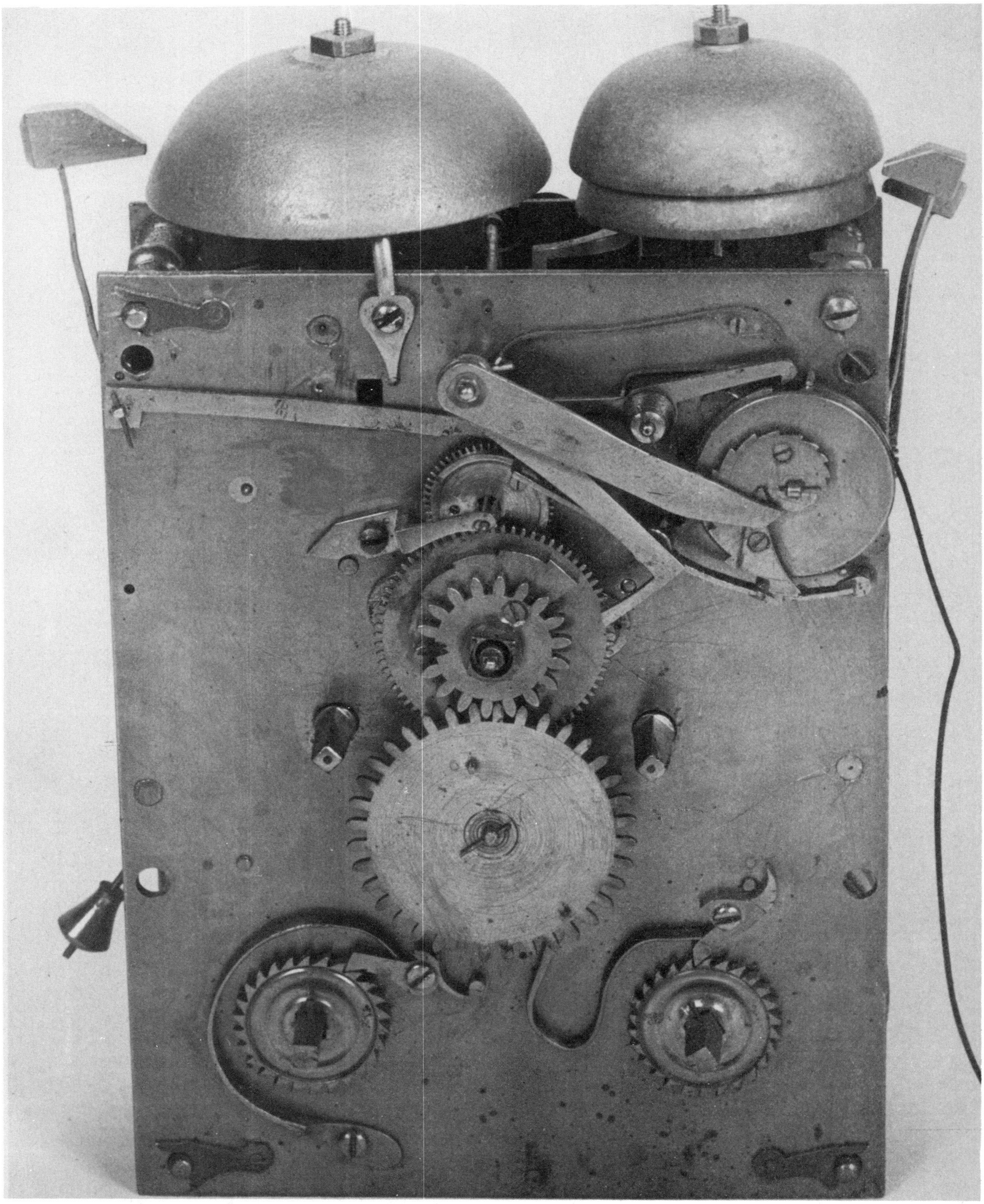

Massy. This transitional clock, which has locking plate striking for the passing hours, also has a completely independent silent pull sounding the hours and quarters without disturbing the locking plate. Although the quarters are determined conventionally by means of a rocking bar, the silent pull hours are achieved by a system almost identical with that used for the quarters in those clocks shown on pages 82-91.

Striking Clock with Pull Quarter on Three Bells

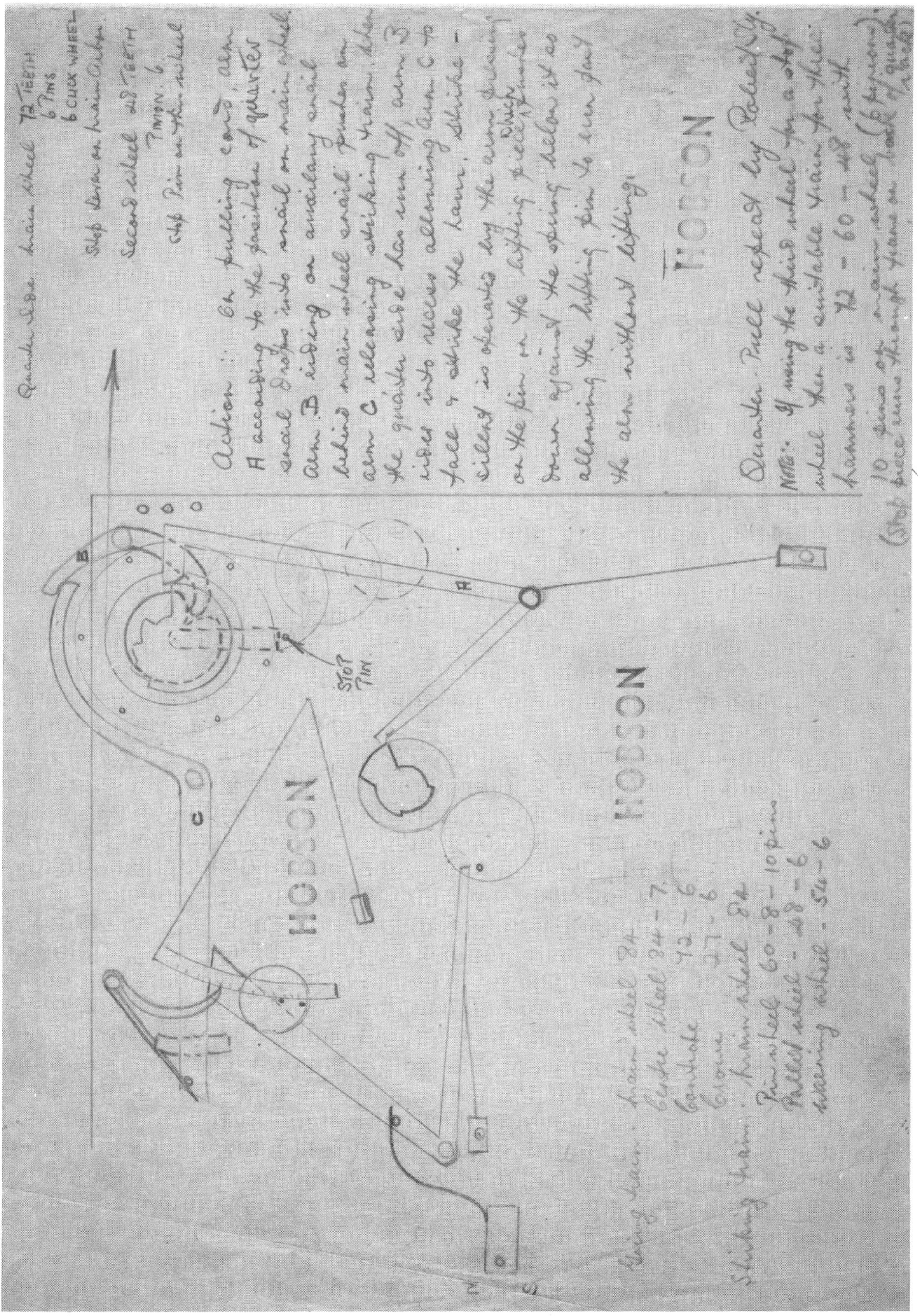

Robert Sly. Details of an arrangement employing conventional pull quarter work but having an unusual arrangement for letting off the striking work. Although it is nowhere stated in the sketch, this clock apparently employs an inside rack, a highly unusual arrangement at this late date.

Striking Clock with Pull Quarter on Six Bells

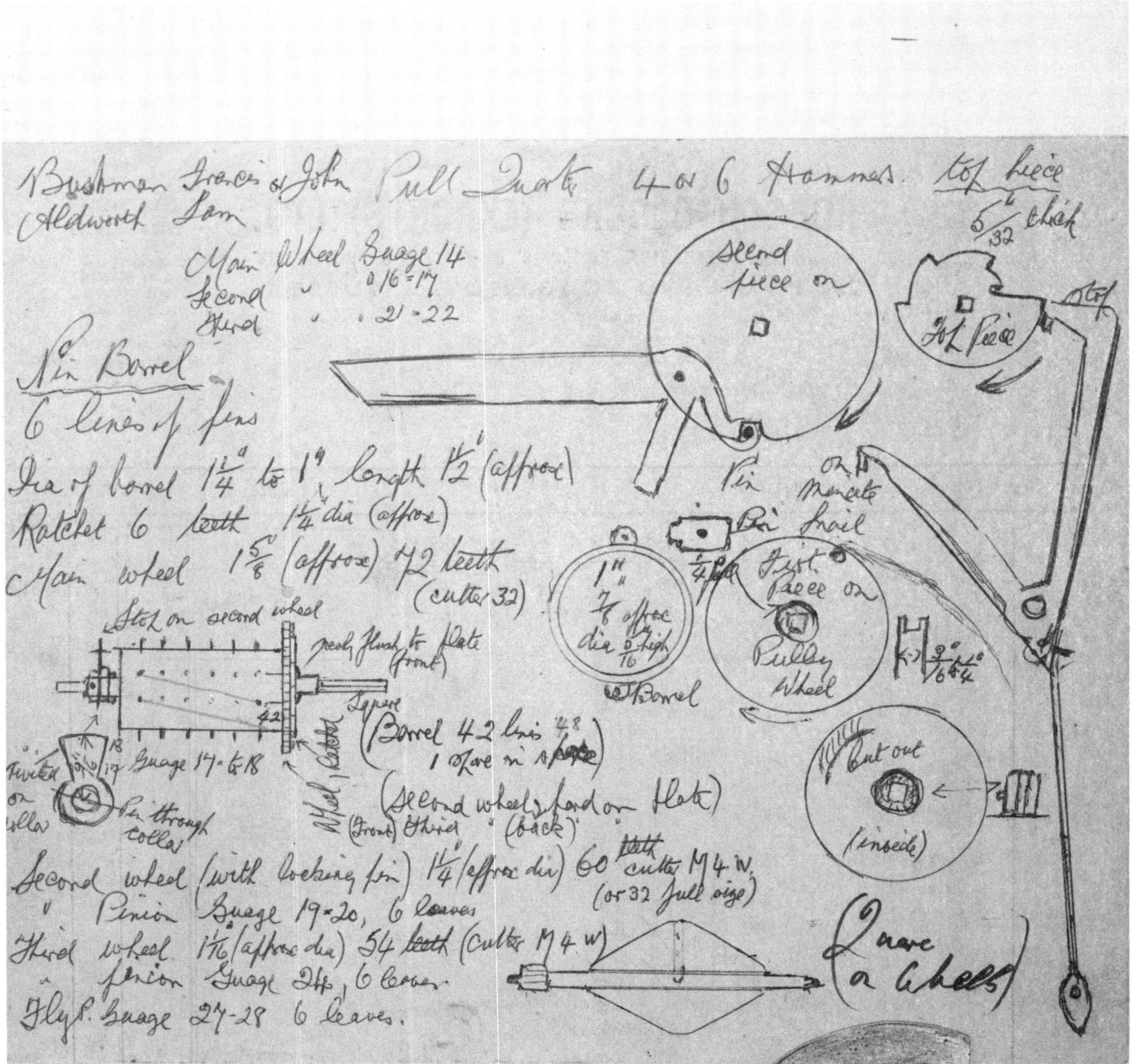

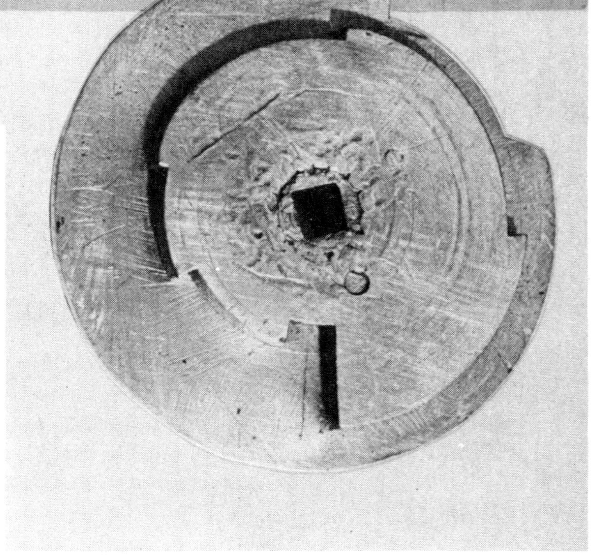

John & Francis Bushman, also Sam Aldworth. Details of pull quarter rack and snail, hour striking releasing mechanism, spring barrel and pulley and pull quarter train count. Also shown are pull quarter pin barrel with main and click wheels and stop arm. This system became common in the 18th century. The small separate illustration shows a combined quarter snail and hour striking release wheel. The two are sketched separately at top right. See also page 90.

Striking Clock with Pull Quarter on Six Bells

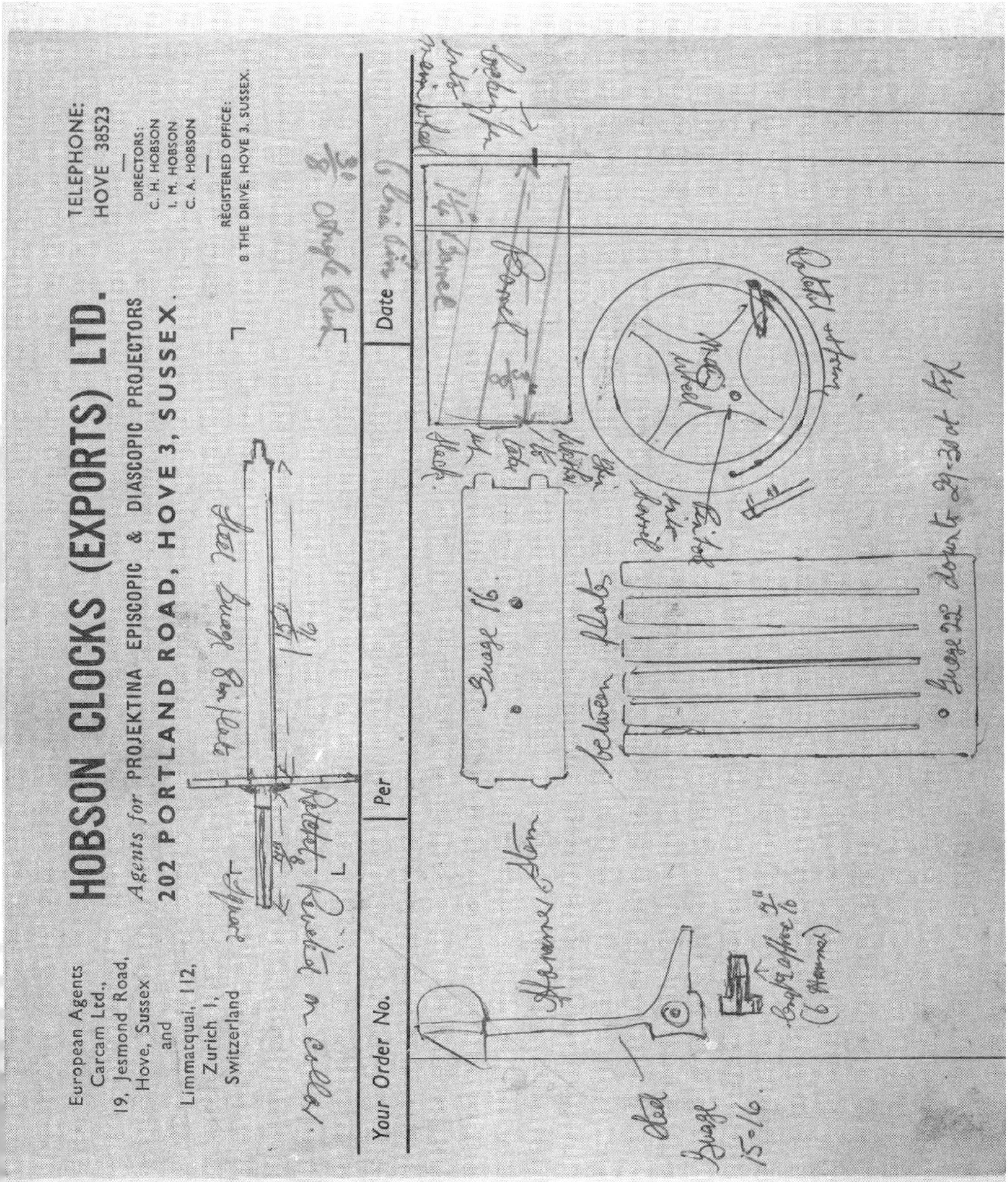

John & Francis Bushman, also Sam Aldworth. Further details taken from the clock on page 86, showing hammer profile, hammer-return springs, main wheel and clickwork and plotting of pins on barrel.

Striking Clock with Pull Quarter on Three Bells

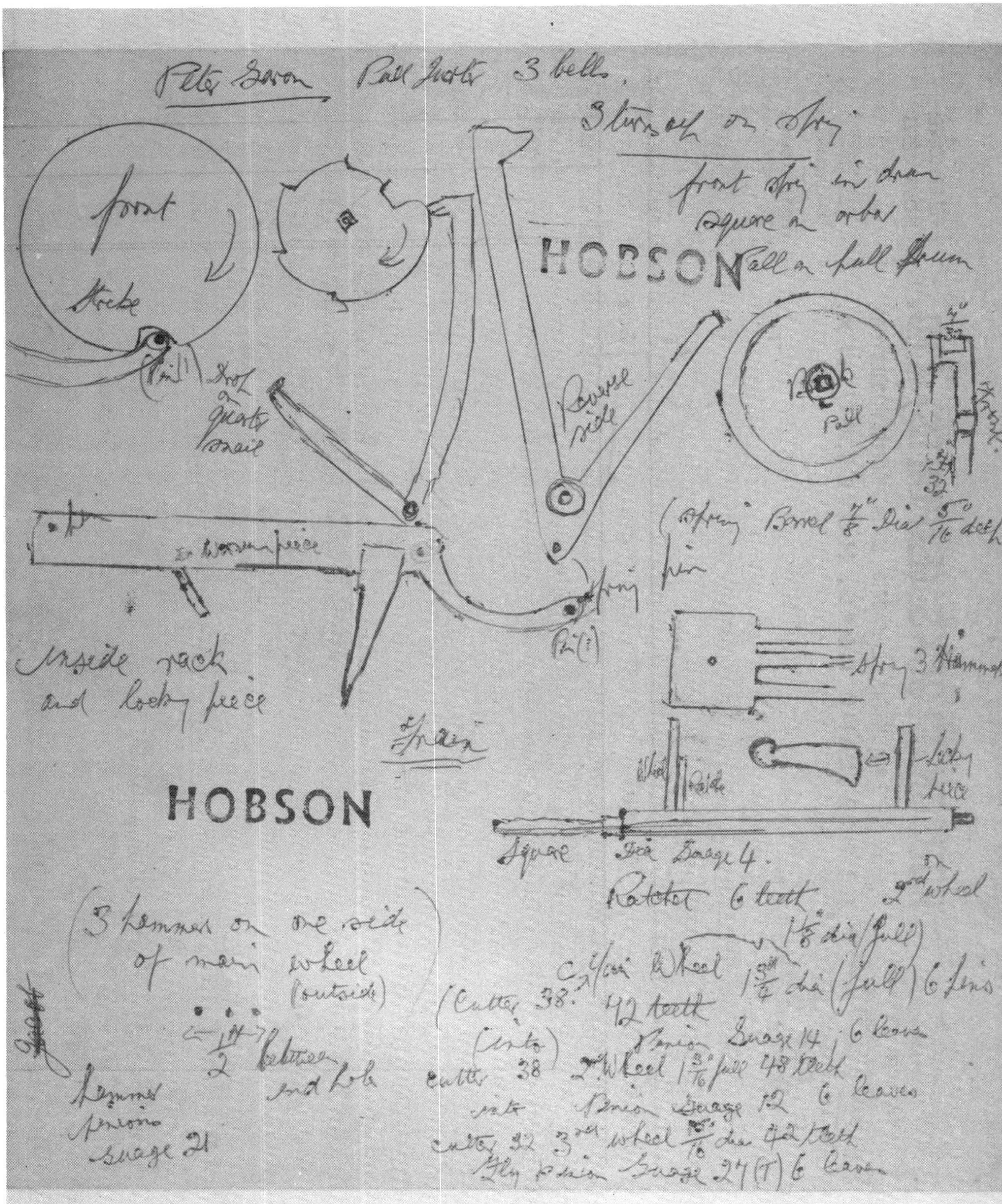

Peter Garon. Details of individual parts used in an arrangement broadly similar to that on page 86, but employing 3 hammers planted between the plates rather than 6 hammers in a hammer block.

Striking Clock with Pull Quarter on Five Bells

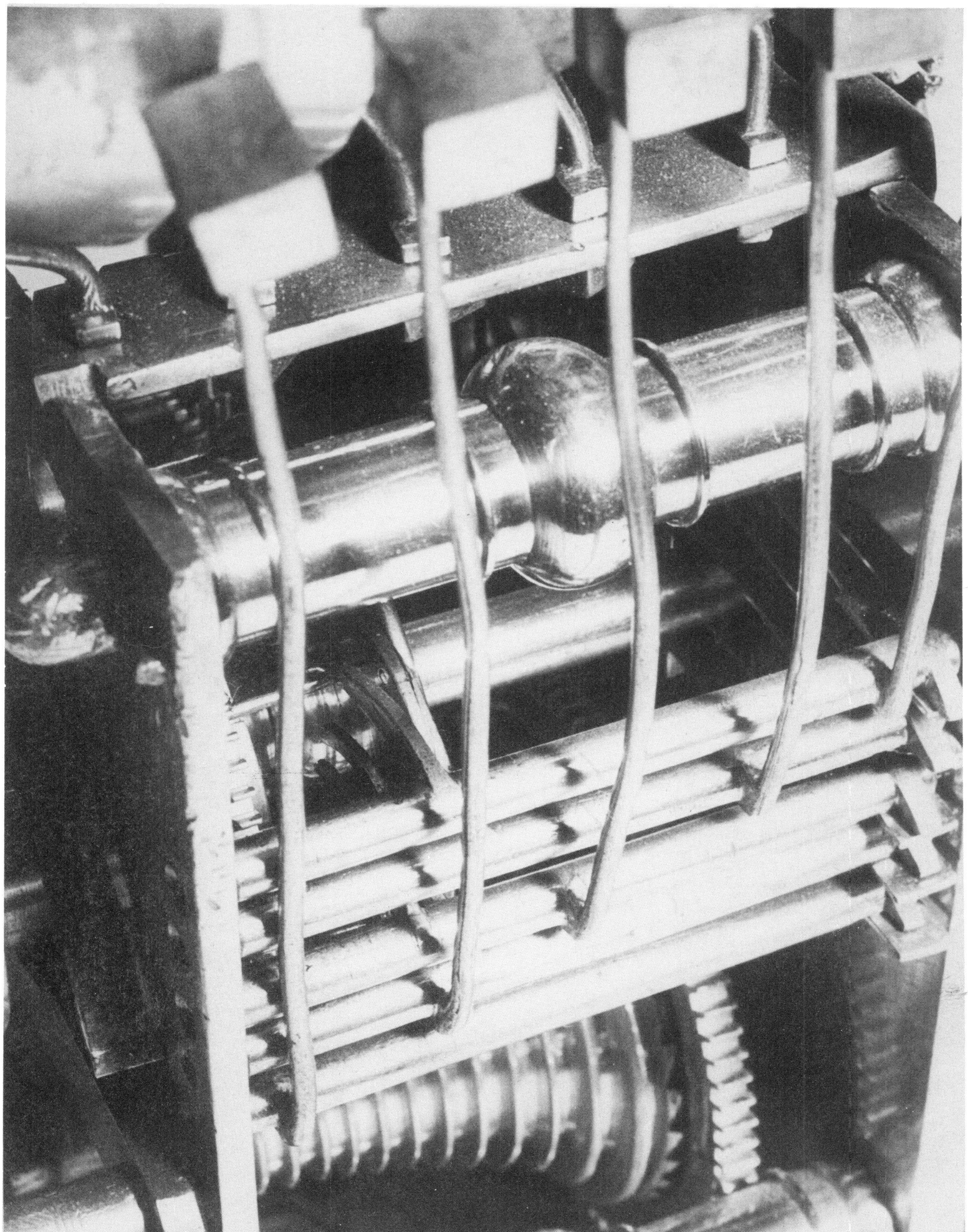

Quare. Method of planting multiple quarter hammers actuated by a pin wheel. A later system employed a pin barrel used in conjunction with hammers mounted in a block.

Striking Clock with Pull Quarter on Three Bells

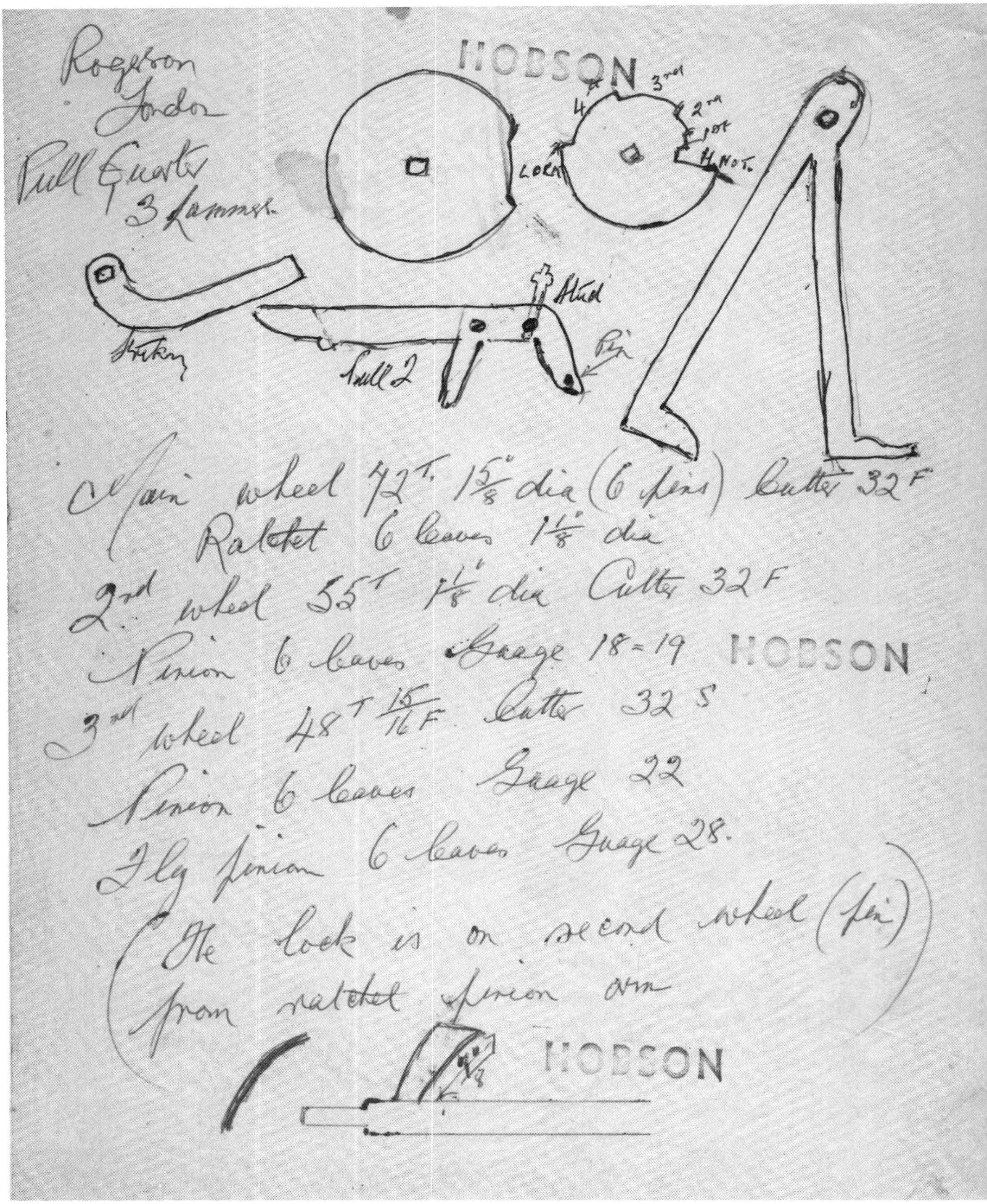

Rogerson. Details of a clock broadly similar to that shown on pages 86 to 88.

Striking Clock with Pull Quarter on Two Bells.

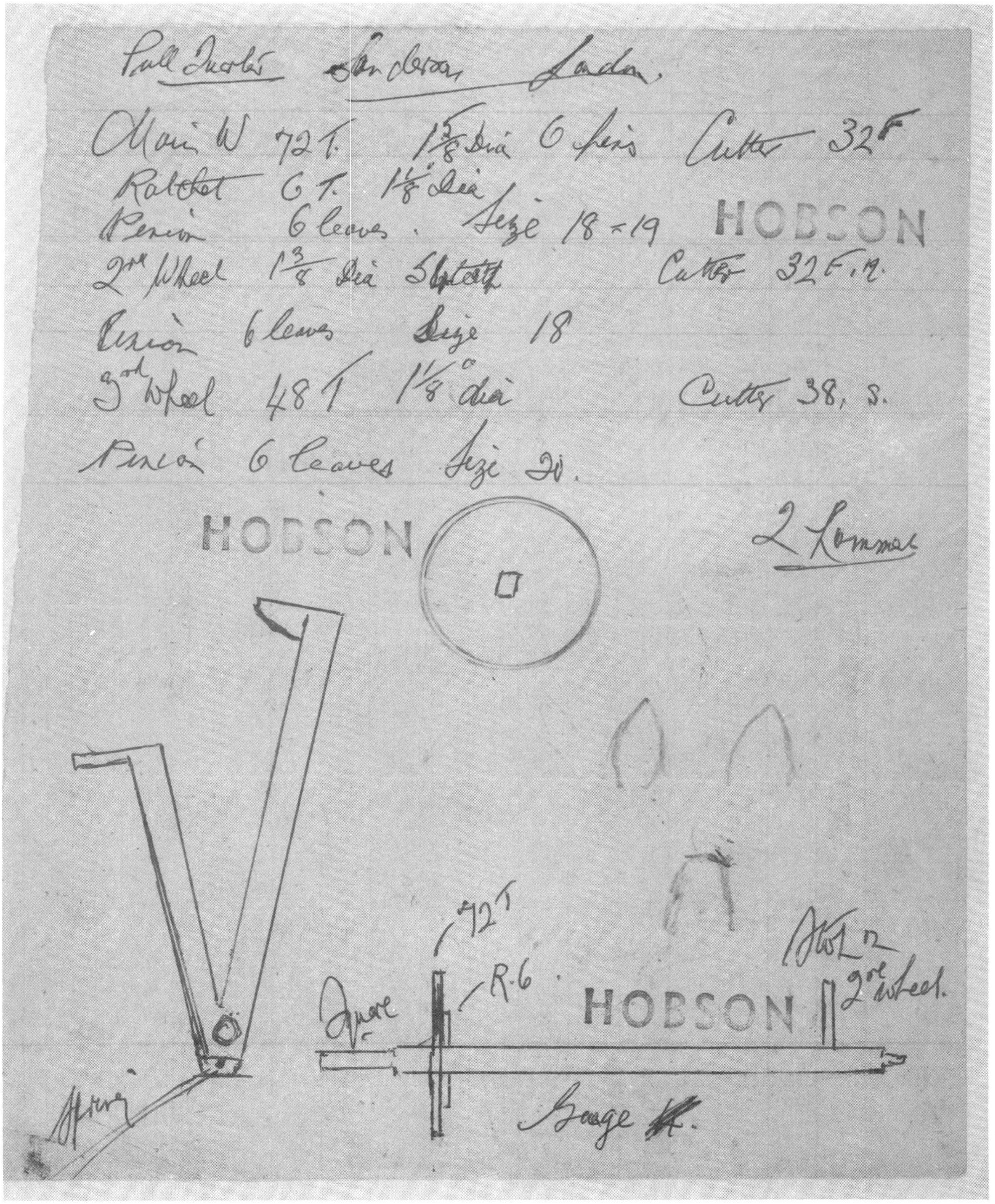

Sanderson. Details of pull quarter work similar to that shown on pages 86-90.

Striking Clock with Pull Quarter on Six Bells

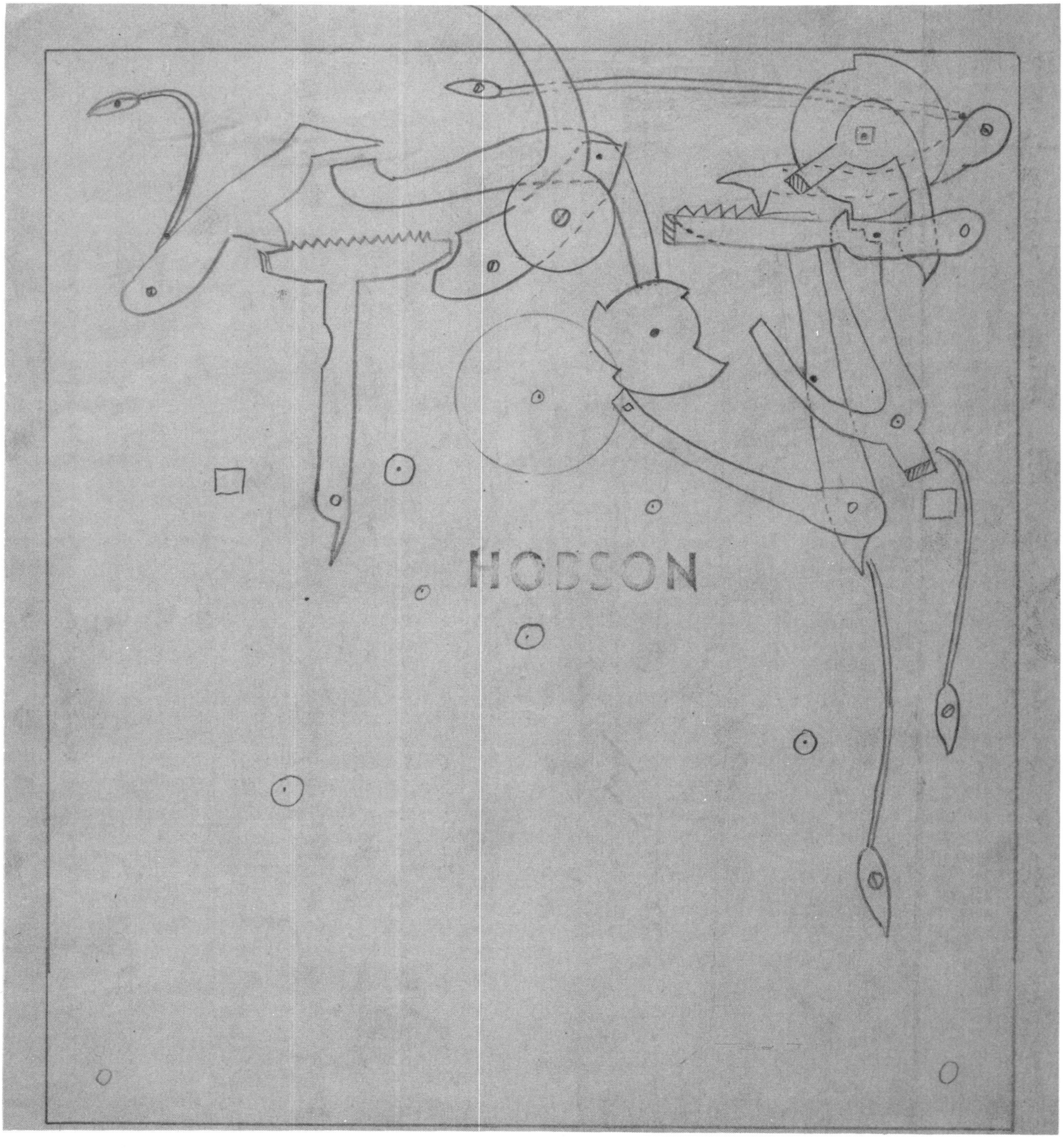

Six bells with pin barrel. Spring box and pulley on back. Hammer frame screwed on back. 10 rows of pins.

Train 70
60 . 7
54 . 6
6

Pinchbeck. Details of under-dial work similar to that shown in page 98, although with minor variations.

Striking Clock with Pull Quarter on Six Bells

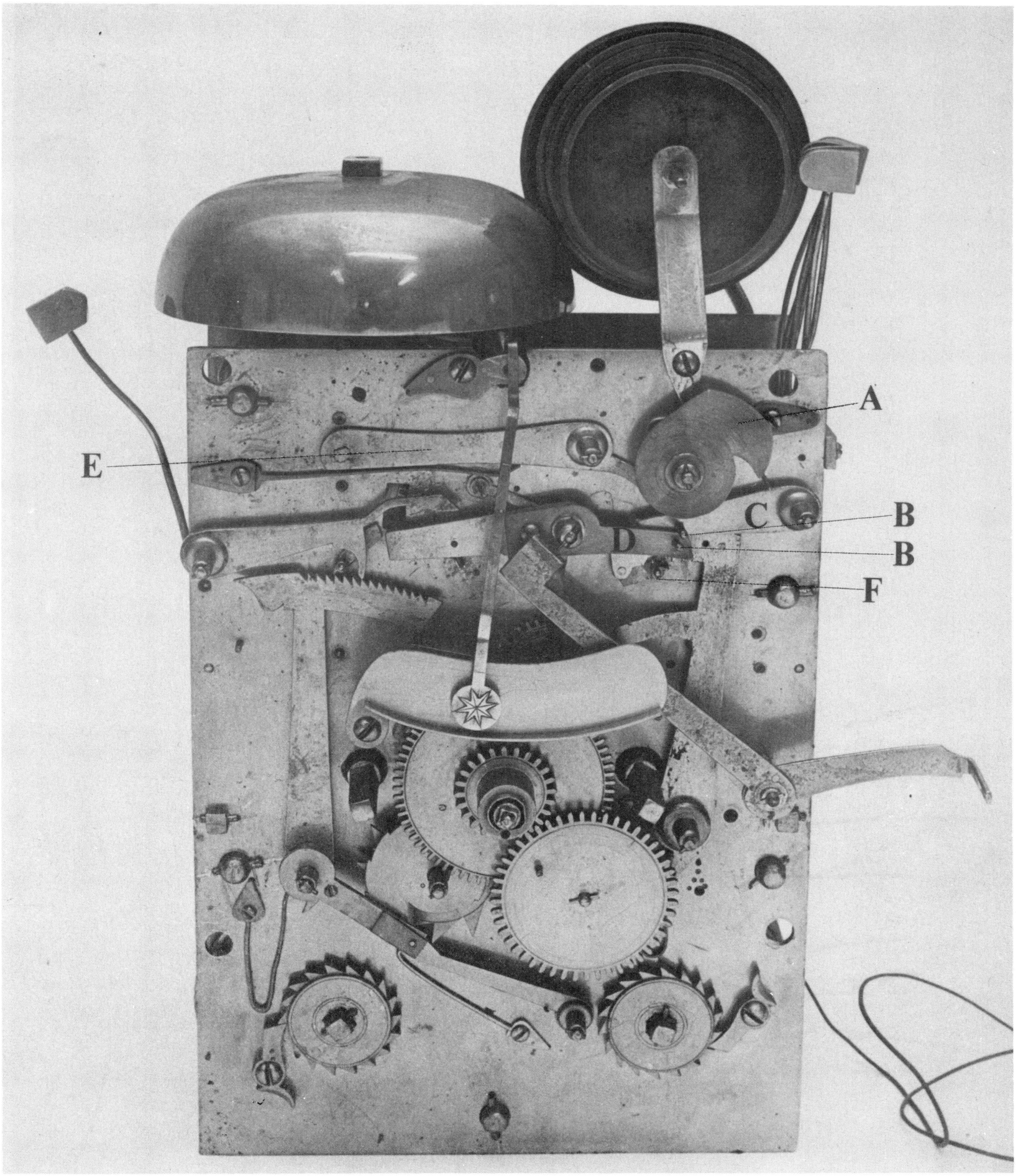

Windmills. On pulling off, and as the wheel A turns clockwise, the beak enters between the two pins BB thereby raising the quarter rack hook C and at the same time depressing the tail of the hour striking set off lever D. This raises the hour rack hook but the train is held on the warn by the lever E. The lever D is immediately released leaving the hour striking train still held on the warn. On release of the pull, the double-toothed quarter rack (see page 98) is gathered by the gathering pallet F. On completion of the gathering of the quarters with the consequent return of the wheel A to its original position the tail of lever E drops once again into its slot in A thereby setting off the hour striking train. The power is supplied by a spring under the pulley on the back plate. See page 100 for the general idea.

Striking Clock with Pull Quarter on Six Bells

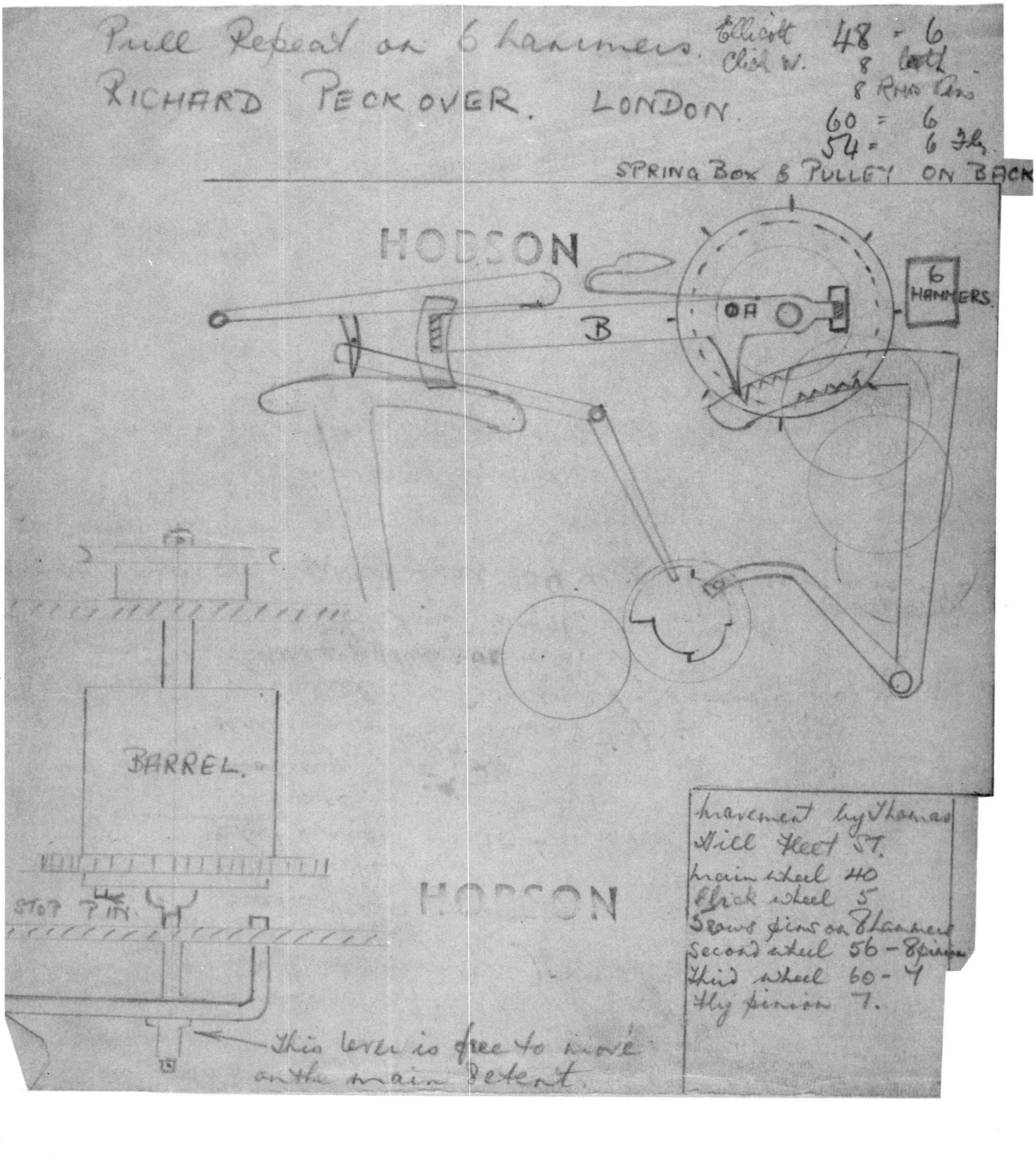

Richard Peckover. View under dial showing pull quarter and hour striking work; also the position of the hammer block in relation to the pin barrel. Sketch at bottom shows plan view of pulley, spring barrel, main wheel and clickwork. Note quarter rack hook ("B" in main sketch) showing the tail passing through the front plate. The note top right referring to Ellicott applies to a similar clock (not shown). See page 96 for photograph of the actual clock.

Striking Clock with Pull Quarter on Six Bells

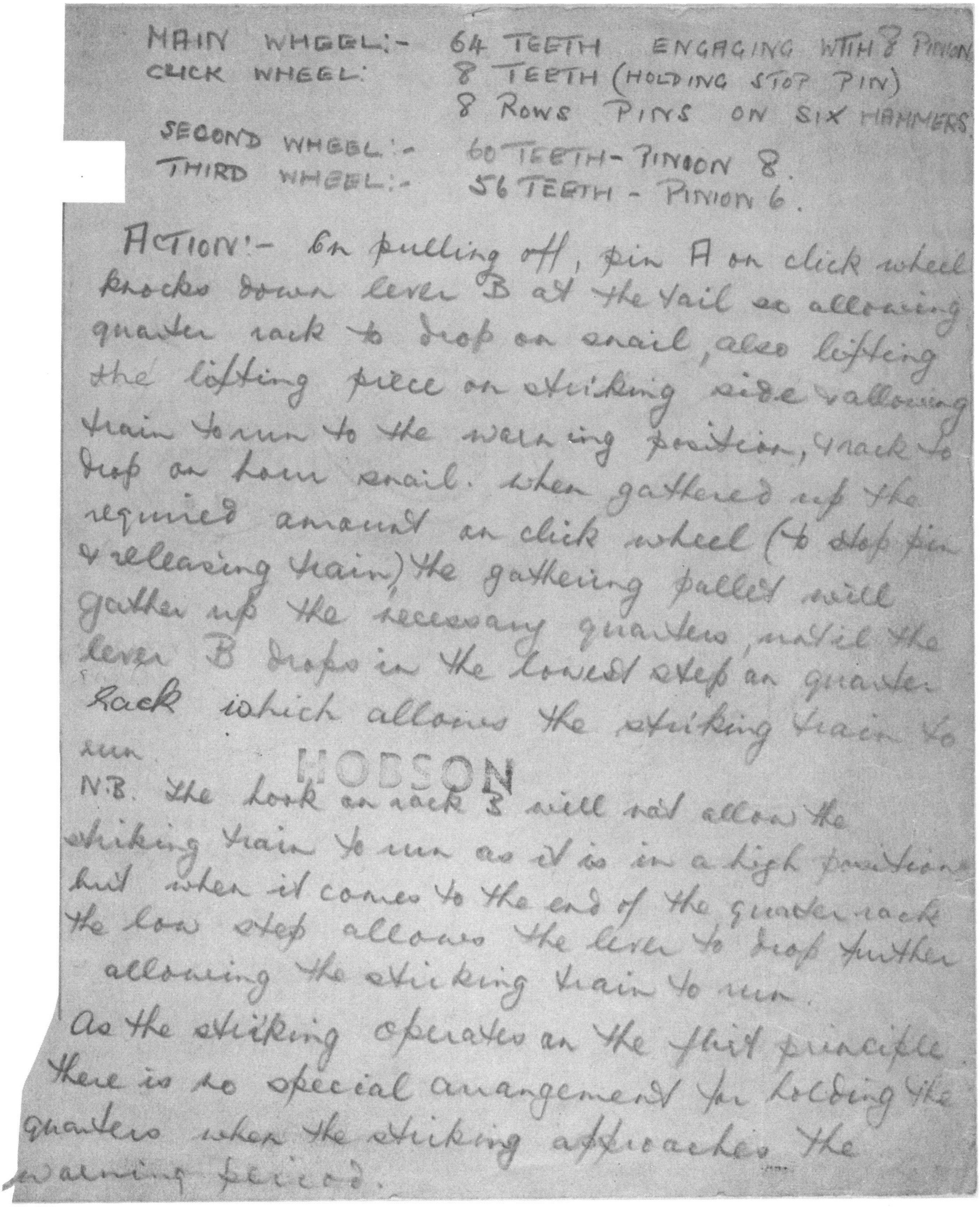

MAIN WHEEL:- 64 TEETH ENGAGING WITH 8 PINION
CLICK WHEEL: 8 TEETH (HOLDING STOP PIN)
8 ROWS PINS ON SIX HAMMERS.
SECOND WHEEL:- 60 TEETH - PINION 8.
THIRD WHEEL:- 56 TEETH - PINION 6.

ACTION:- On pulling off, pin A on click wheel knocks down lever B at the tail so allowing quarter rack to drop on snail, also lifting the lifting piece on striking side allowing train to run to the warning position, & rack to drop on hour snail. When gathered up the required amount on click wheel (to stop pin & releasing train) the gathering pallet will gather up the necessary quarters, until the lever B drops in the lowest step on quarter rack which allows the striking train to run.

HOBSON

N.B. The hook on rack B will not allow the striking train to run as it is in a high position but when it comes to the end of the quarter rack the low step allows the lever to drop further allowing the striking train to run.

As the striking operates on the flirt principle there is no special arrangement for holding the quarters when the striking approaches the warning period.

Richard Peckover. Pull quarter train and description of action in the clock shown opposite on page 94. See page 96 for photograph of the actual clock.

Striking Clocks each with Pull Quarter on Six Bells

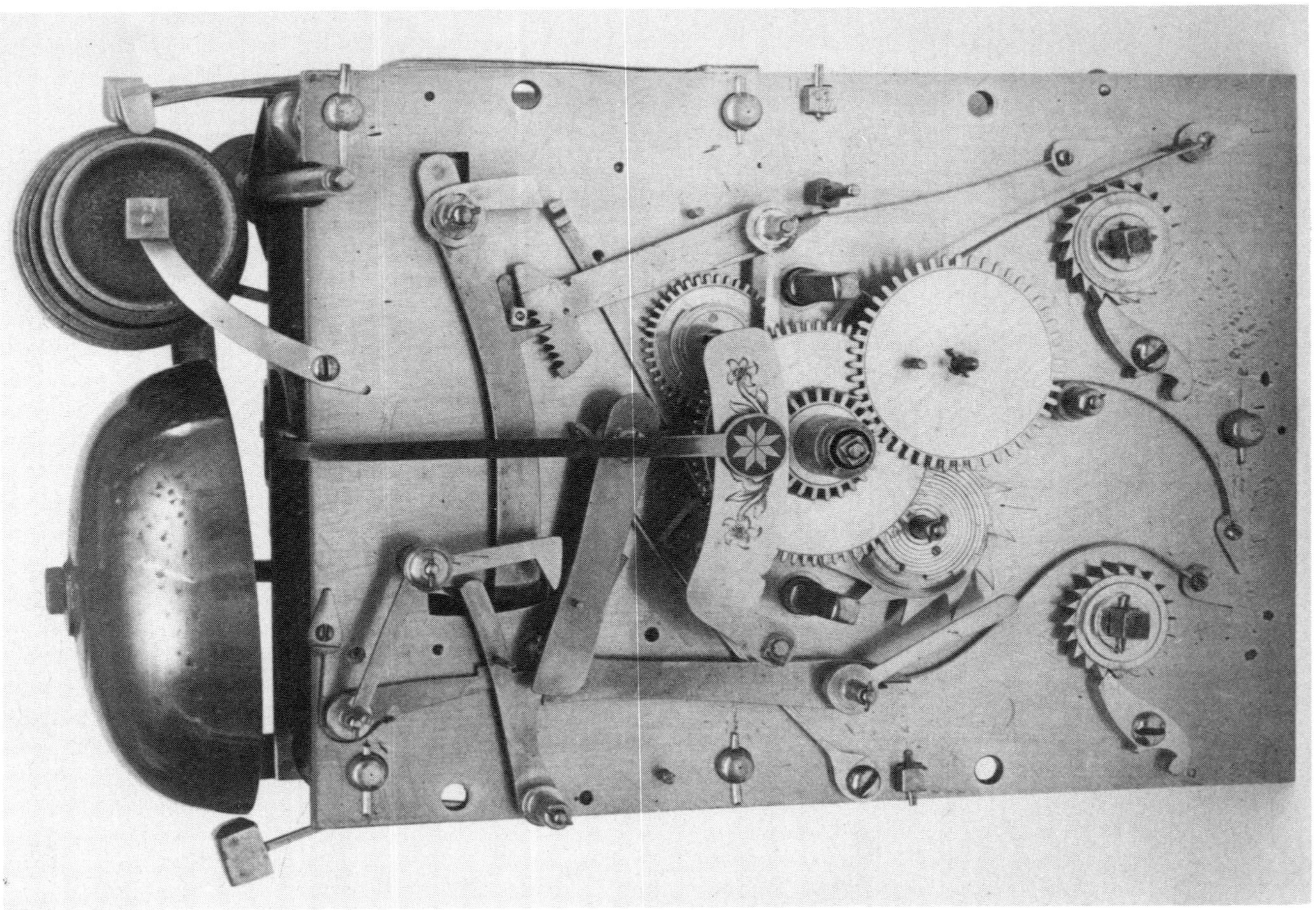

Hewitt. This clock is similar in operation to the one by Peckover already illustrated on pages 94-96.

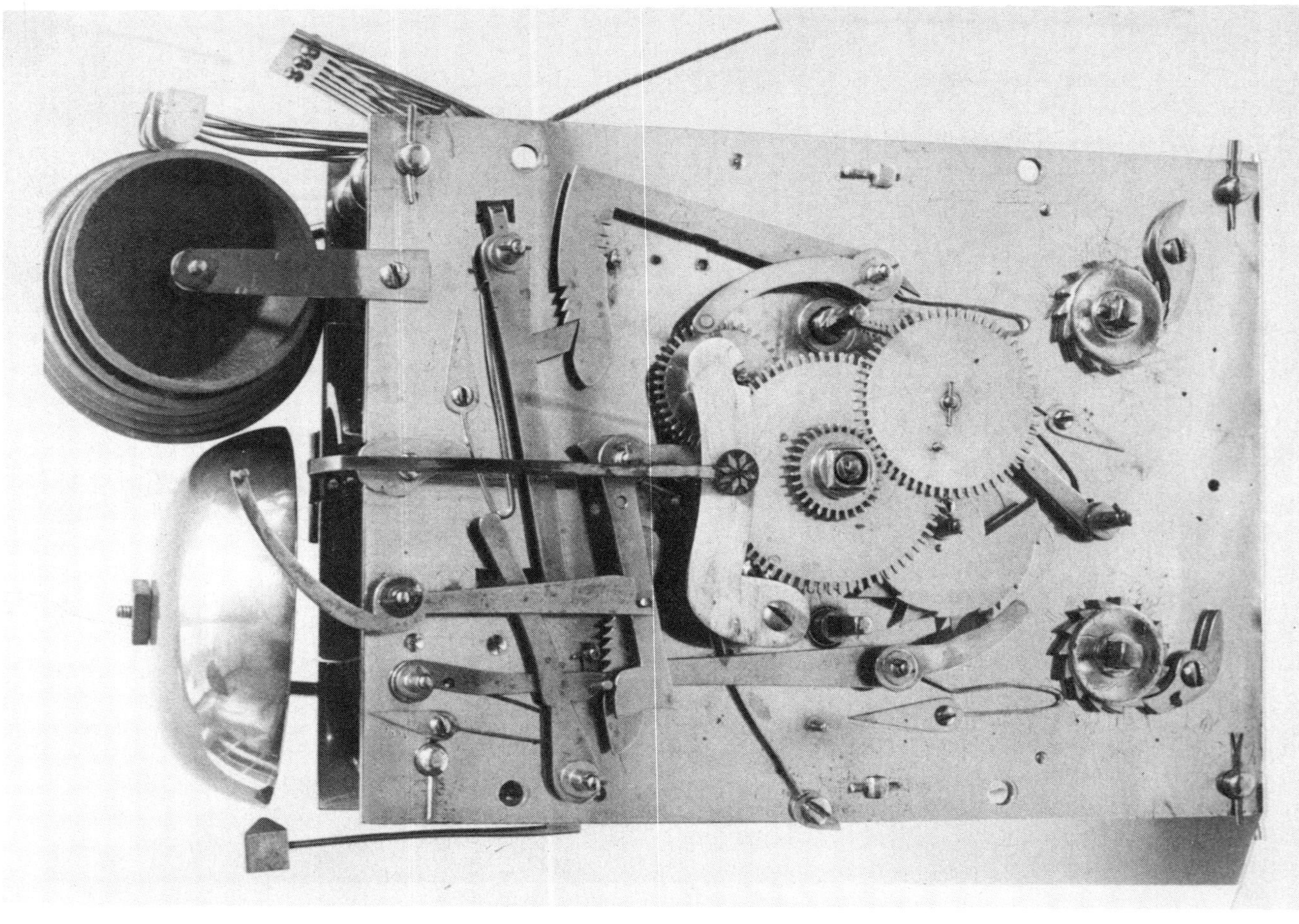

Peckover. This clock is described on pages 94 and 95.

Striking Clock with Pull Quarter probably on Six Bells

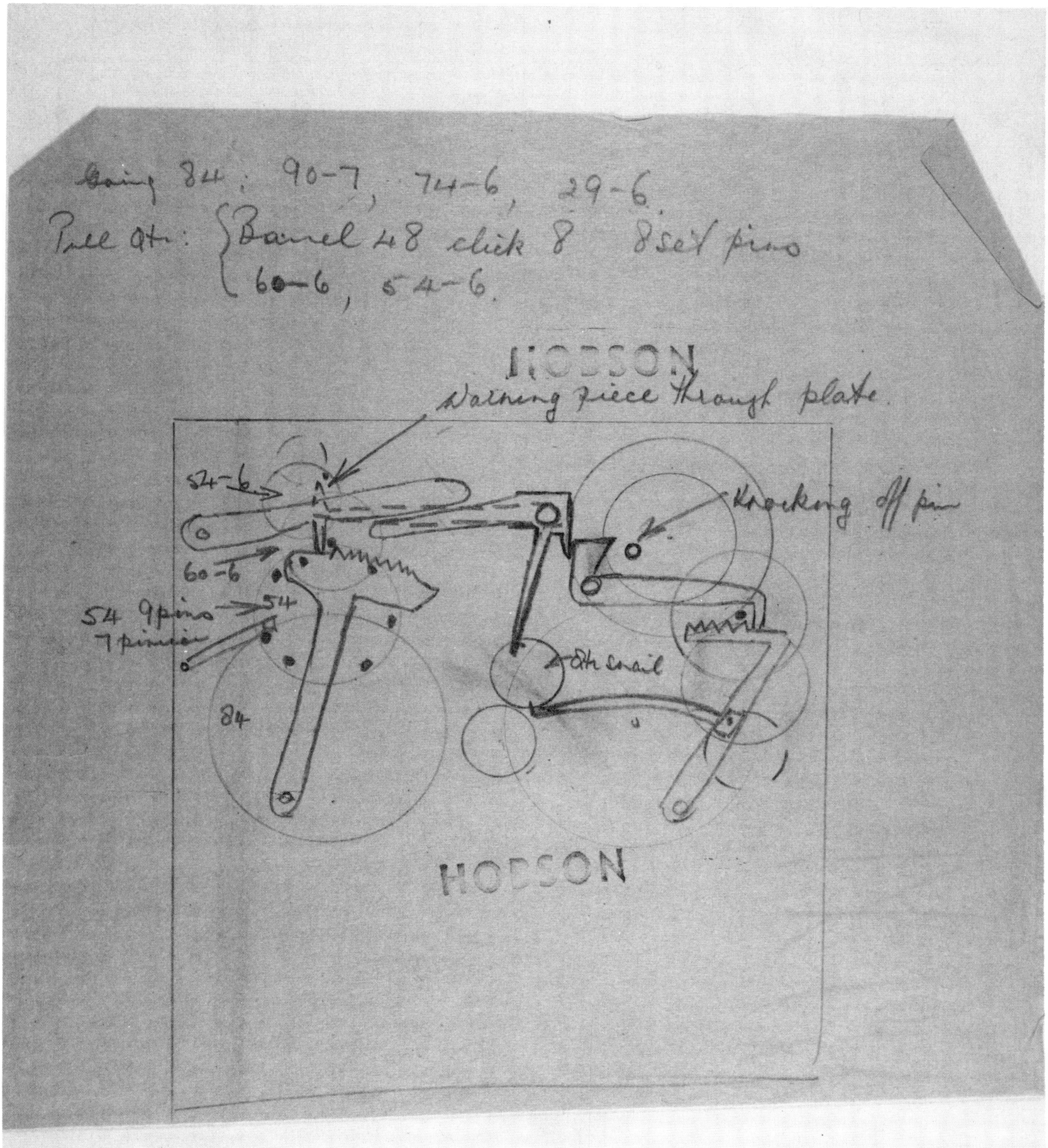

Unidentified. On pulling the cord, which is held on a pulley on the back plate, the brass wheel carrying the knocking off pin turns clockwise soon striking the tail of the quarter rack hook. This releases the quarter rack and also warns the hour striking train. When the pull is released the quarter rack is gathered, eventually releasing the hour striking train in the same manner as described on page 95.

Striking Clock with Pull Quarter probably on Six Bells

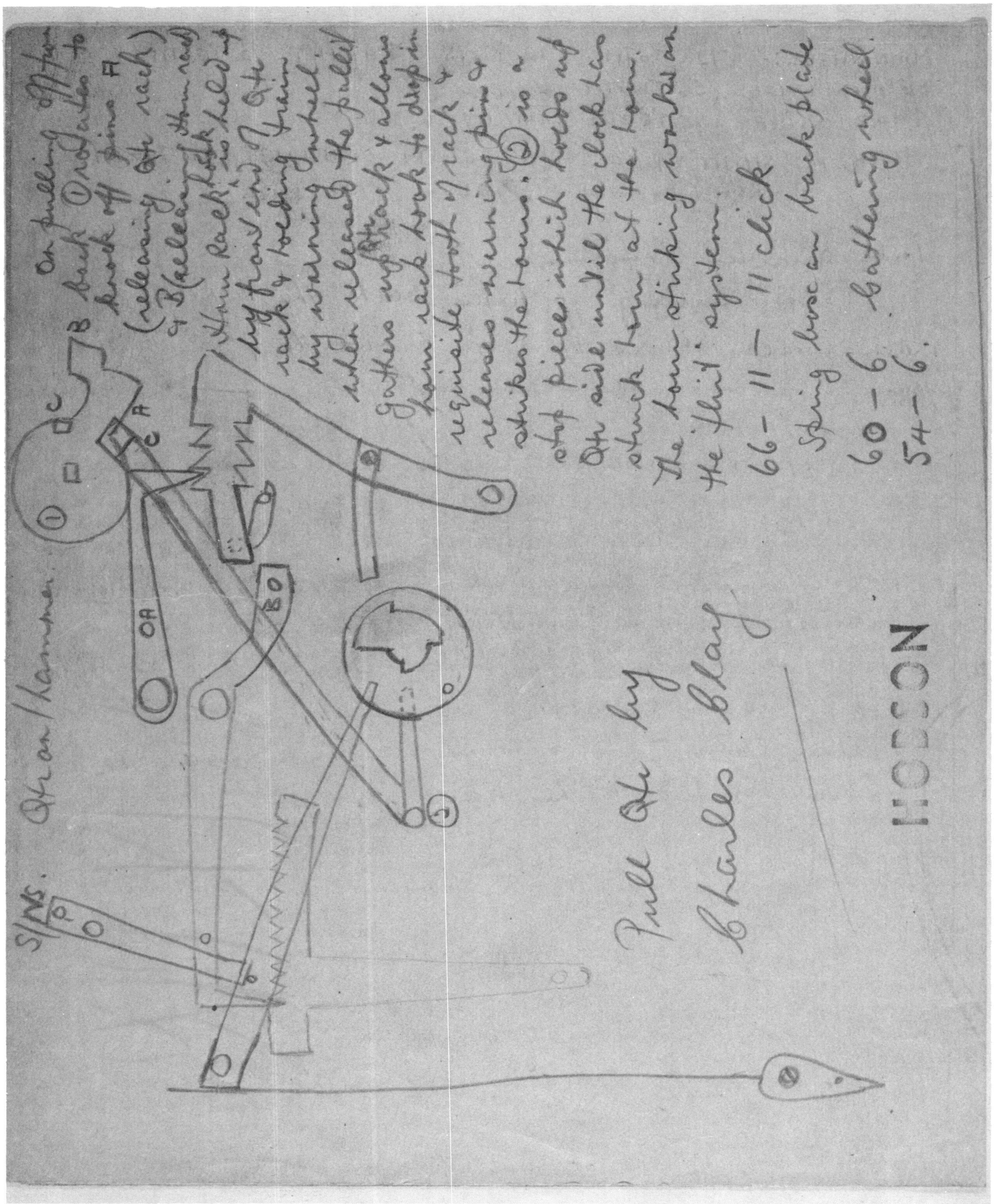

Charles Clay. The action here is broadly similar to that discussed on page 97.

Striking Clock with Pull Quarter on a Single Bell

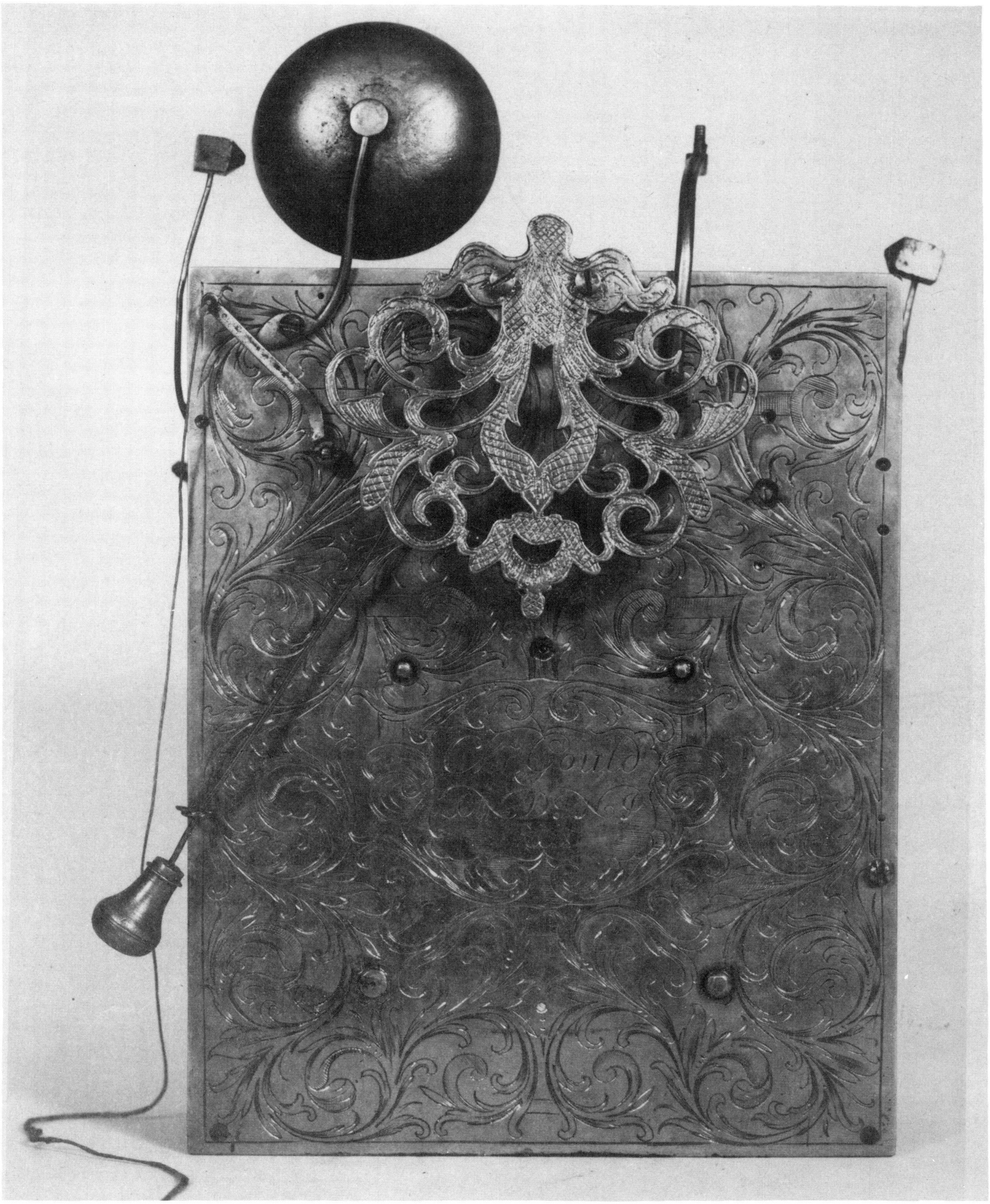

Christopher Gould. The back plate and pull off lever of the clock discussed on page 83.

Striking Clock with Pull Quarter on Six Bells

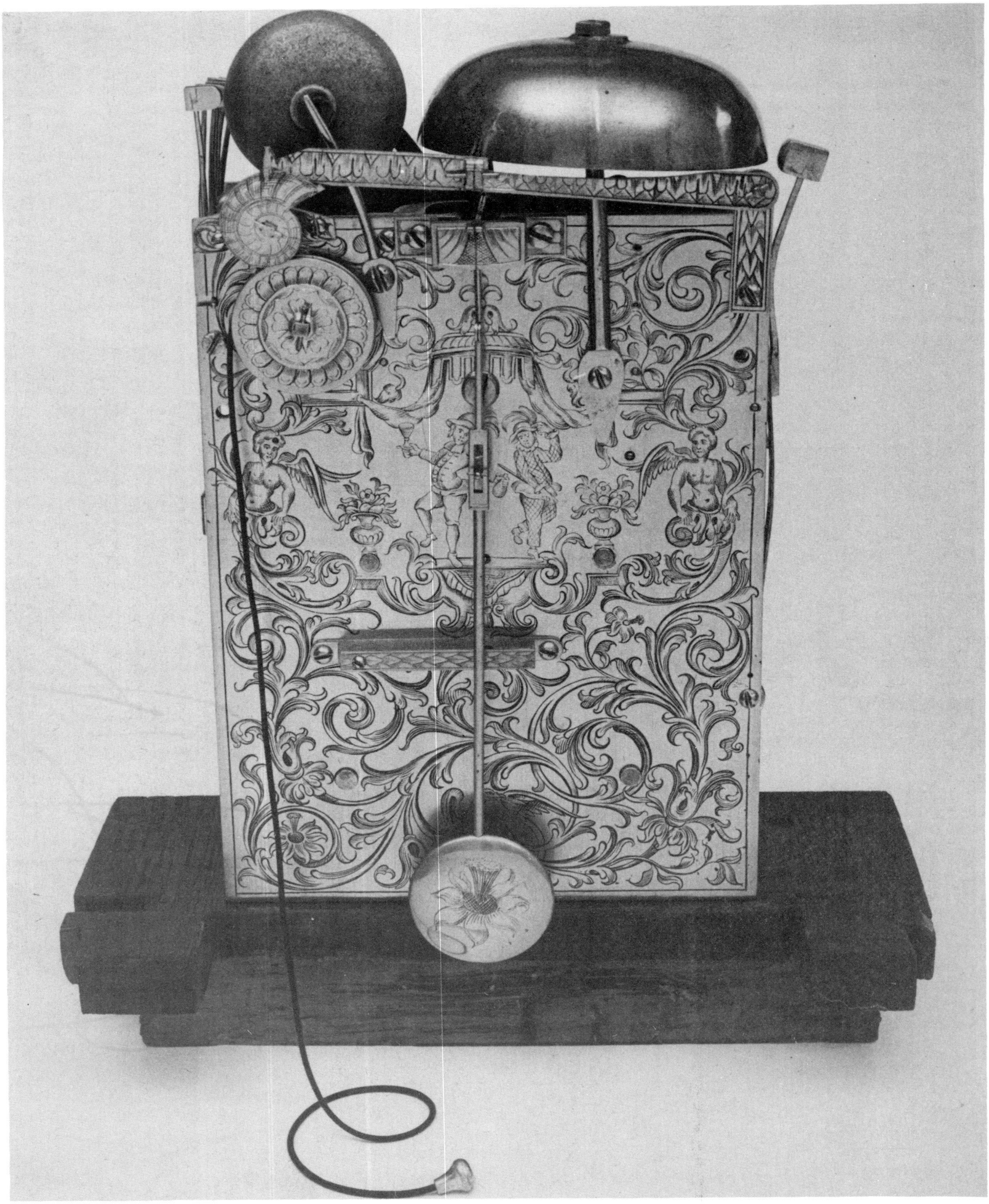

Hewitt. Well engraved back plate with typical matching pulley covering pull quarter barrel and spring.

Original pulley box in bottom of the Case of the Clock shown on page 96 left

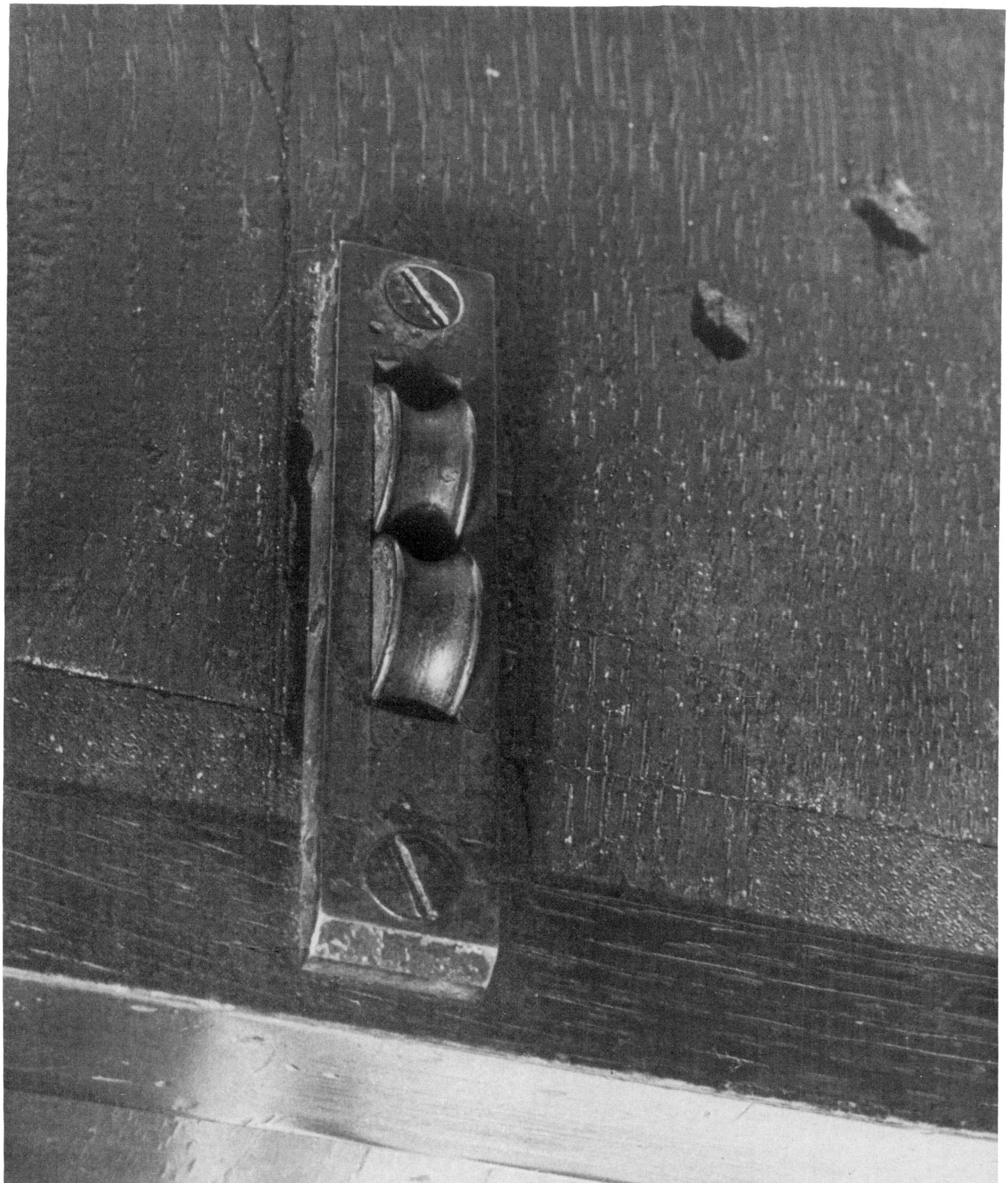

Peckover. Pulley boxes were never very much used although the idea seems attractive at first. Almost certainly makers soon discovered that once the slightest wear had taken place the cord always ended up between the pulley and its frame. This trouble was most noticeable with bottom pulley boxes which involve two wheels. (Refer also to page 47 and to Lee *Knibb*, plate 84).

NOTES

NOTES